COUNSELOR
INTERN'S
HANDBOOK

Michael Cerullo
61 William Reynolds
Exeter, RI 02822
295-0870

THE COUNSELOR INTERN'S HANDBOOK

Christopher Faiver
Sheri Eisengart
Ronald Colonna
John Carroll University

Brooks/Cole Publishing Company
Pacific Grove, California

I(T)P ™ The trademark ITP is used under license.

A CLAIREMONT BOOK

Brooks/Cole Publishing Company
A Division of Wadsworth, Inc.

Printed in the United States of America
10   9   8   7   6   5   4

Library of Congress Cataloging-in-Publication Data

Faiver, Christopher
        The counselor intern's handbook / Christopher Faiver, Sheri
Eisengart, Ronald Colonna.
        p.   cm.
        Includes bibliographical references and index.
        ISBN 0-534-24870-5
        1. Counseling—Study and teaching (Internship)—Handbooks,
manuals, etc.   I. Esengart, Sheri.   II. Colonna, Ronald.
        III. Title.
BF637.C6F32   1994
158'.3'07155—dc20   94-18836
                                             CIP

Sponsoring Editor: *Claire Verduin*
Marketing Representative: *Thomas L. Braden*
Editorial Associate: *Gay C. Bond*
Production Editor: *Penelope Sky*
Production Assistant: *Tessa A. McGlasson*
Manuscript Editor: *Kathleen Pruno*
Permissions Editor: *Lillian Campobasso*
Interior and Cover Design: *Vernon T. Boes*
Indexer: *Do Mi Stauber*
Typesetting: *Bookends Typesetting*
Printing and Binding: *Malloy Lithographing, Inc.*

*To our families,
friends, mentors, and students*

CONTENTS

PREFACE

This basic guide is designed to assist counseling students and others in the helping professions through the entire experience of internship, often the last requirement of a degree program. We include overviews of basic treatment modalities, psychological testing, and psychopharmacology, as well as chapters on the clinical interview and ethical considerations, subjects that are particularly relevant to interns. Our fields are counseling and psychology, yet students in related areas of the behavioral sciences may find the book helpful.

Among us we have supervised field placement students in a variety of settings for more than 40 years, so we are able to maintain a practical perspective in this book. Our varied backgrounds include teaching, administration, and clinical work; one of us recently completed the internship. We encourage students to review what they have learned while they progress to new material. We explain the components of internship in detail, and urge students to use them with enthusiasm as well as caution. It is important to leave no stone unturned during the internship process, but we provide only the stones, respecting the student's responsibility for the turning.

Acknowledgments

Some of the information we offer emerged from institutions involved in providing field experiences to interns; in particular, we thank the committees of the John Carroll University Counseling and Human Services Program and the Youngstown State University Department of Counseling. A number of students and colleagues contributed valuable assistance while we were writing this book. They include Christy Berardinelli, John Bufford, Florence Ciccone, Seth Eisengart, Elliott Ingersoll, Fabian Newman, John Ropar, Nancy Salkin, and Keith Smedi. We also appreciate the efforts of our reviewers: Sari H. Dworkin, California State University at Fresno; Daisy B. Ellington, Wayne State University; Peter Maynard, University of Rhode Island; Beverly Palmer, California

State University, Dominguez Hills; and Frederick Sweitzer, University of Hartford.

The staff at Brooks/Cole was enormously helpful. This book would not have been written without the encouragement and support of Tom Braden, Claire Verduin, and Gay Bond. We also appreciate the efforts of Penelope Sky, Kathy Pruno, and Carline Haga.

We encourage your comments. Please write to us at Brooks/Cole Publishing Company, 511 Forest Lodge Road, Pacific Grove, California 93950.

Christopher Faiver
Sheri Eisengart
Ronald Colonna

THE
COUNSELOR
INTERN'S
HANDBOOK

CHAPTER ONE

GETTING STARTED

Your internship is generally the culmination of the academic sequence leading to your degree in counseling. It is an exciting, challenging time, and many students anticipate it with mixed emotions. Internship students often find themselves facing unfamiliar situations, engaging in intense encounters, and processing powerful feelings, which all lead to increased introspection, personal reassessment, change, and growth. The internship may be your first opportunity to interact independently with real clients in a professional capacity as a counselor, as well as your first experience in submitting those interactions to scrutiny during a formal supervisory hour. The material presented in this book is intended to be used as a step-by-step guide that will take some of the anxiety out of your internship and allow you to relax and enjoy the many rewarding, wonderful moments you are about to experience as a counselor intern.

Ideally, your internship should provide you with a supportive, structured learning environment for acquiring clinical experience and practical on-the-job training. You will be called on to synthesize material from previous coursework, to utilize theories and techniques, and to begin to develop a personal and professional style of relating effectively to clients, clients' families, agency staff members, and other mental health professionals.

During your counseling internship, you will work under the direct supervision of a licensed counselor, social worker, psychologist, or psychiatrist. You will meet for regular supervisory sessions to review your internship experience, as specified by your state licensure/certification board. Ideally, your field placement supervisor should be an experienced clinician who will provide a flexible learning experience tailored to meet your individual needs by giving you encouragement and ongoing assessments of your strengths and weaknesses as well as asking for your feedback. The actual number of hours you spend in direct contact with clients, as well as the total number of supervisory hours, should always fulfill state licensure/certification requirements. For example, the Ohio State Counselor and Social Worker Board specifies that one quarter of the internship hours be spent in direct client contact and that the intern should have one hour of supervision for every 20 hours of work. That would mean that a graduate student fulfilling a 600-hour internship would need 150 hours of client contact and 30 hours of supervision to have the internship qualify toward state licensure.

In addition to on-site experience and supervision, you will most likely participate in on-campus class meetings with other interns to discuss your placement experiences. Most students appreciate this group time with their peers to share their feelings, their frustrations, and their accomplishments, and find the support this group offers to be especially helpful during their internships. You may also meet with a supervisor from your counseling program (usually one of the department or program professors) for further processing of your individual experiences. You will typically be required to keep a detailed log or journal of daily on-site activities and to prepare two or more formal case studies for class presentation. Additional research projects or reading relevant to counseling may be assigned, as well. The agency supervisor and the program supervisor will evaluate your performance and potential as a counselor, and ideally they will discuss their evaluations with you midway through the internship and again at completion of the experience.

SELECTING YOUR INTERNSHIP SITE

Selecting your placement site is the first, and perhaps the most important, step in your internship experience. The selection process ideally involves integrating four factors:

1. your own interests and needs;
2. the field placement guidelines of your university counseling or human services program;
3. the state requirements for on-the-job experience for professional counseling licensure/certification; and
4. the didactic and experiential opportunities afforded at the placement site.

These four factors are interdependent, and the selection of a placement site is a dynamic process of exploring and matching these various criteria to find a good fit. Some counseling and human services programs have formal placement relationships with preapproved agencies, which may limit selection to a certain extent, but which may also help to provide a satisfactory field experience. Other programs delegate most of the responsibility for securing field placements to the individual student. In either case, you should first try to determine your own interests, needs, and expectations when you begin the process of choosing a placement site.

Your Interests and Needs

You may already have a fairly good idea about what types of people you do, or do not, enjoy working with (for example, children, adolescents, adults, or the elderly) and what kinds of problems you would, or would not, like to deal with (for example, substance abuse, child and family concerns, career counseling, or mental health issues such as depression and anxiety). Knowledge about your own preferences helps give you some direction when you begin looking for an internship site, because you can limit your choices to those places where you are certain that you have a keen interest in agency clients and services. You may also opt to use your internship as an opportunity to try something new, to deepen your self-awareness and to enlarge your scope of experience. For example, if you have been working with abused children but have not had any helping interactions with chemically dependent adults, your internship may provide a chance to obtain practice in a different area.

Agencies can have client types and services that are either homogeneous (such as drug and alcohol treatment centers or child welfare agencies) or heterogeneous (such as community mental health centers or psychiatric hospital units). If you are unsure about your interests, or if you do not have a preference for working with one type of client or one particular type of problem, then you may do well to consider heterogeneous internship sites, which will offer you a wider range of experience.

In beginning the selection process, you should also consider whether an inpatient or an outpatient setting would satisfy more of your needs and interests. The hospital inpatient setting will typically be fast-paced and high-pressured, with a rapid client turnover. If you select a psychiatric inpatient unit for your internship, you will most likely come into contact with a great many clients, most of whom are in acute distress and are manifesting serious psychopathology. Working with the psychiatric inpatient is similar to crisis intervention in that your counseling relationship and interventions serve to help the client stabilize and return to a prior level of functioning. The hospital unit affords you the opportunity to interact with other health and mental health professionals as a member of an interdisciplinary treatment team, as well as to learn to relate to clients who have complicated psychiatric disorders, and these experiences are valuable and interesting. However, the intensity of the emotional upset and the extent of the problematic behaviors encountered in a hospital inpatient unit may not be suitable for all counseling students.

Long-term residential treatment facilities provide clients who generally have chronic, rather than acute, problems or difficult management issues and require a more structured or supportive environment. One advantage of this type of internship is the extended counseling relationship that you may develop with your clients, who are likely to remain at the facility throughout your placement. In addition, you may have special opportunities to observe, learn about, and interact with your clients as they go about their daily activities. An internship at a long-term treatment center for children or adolescents can be particularly rewarding in these respects.

If you choose an outpatient setting, your internship experience will vary according to the particular agency. Some agencies may offer you the chance to work as a member of an interdisciplinary team and to participate in treatment conferences, whereas at other agencies you may be interacting only with your supervisor to discuss cases. Most community mental health centers are mandated by law to provide services in such areas as intake assessment, emergency care, consultation and education, research and evaluation, individual and group counseling, and after-care planning. The outpatient setting will usually, but not always, serve clients who are less distressed and less acute than those in the inpatient unit. As an intern in an outpatient center, you may have more opportunities to use a variety of counseling techniques, because the clients are generally higher functioning and are able to cope more effectively with daily living tasks. However, if you work in the intake or emergency department of the community mental health center, you may be dealing with highly upset clients or clients in crisis, who may require immediate hospitalization. In addition, many of your clients may have chronic problems or be "repeaters" who require continued support and management.

Another aspect of your internship needs may involve financial considerations. At some agencies interns are paid for their services; however, these sites will be limited and therefore your choices will be somewhat restricted if you require financial compensation during your internship. Unfortunately, a great many excellent internships are unpaid positions. We have found that some agencies will hire graduate counseling students to perform a paid job, such as case manager or mental health technician, and then allow the student to set aside a designated number of unpaid hours specified as internship hours, when the student assumes tasks and responsibilities relevant to the internship. Some students find this arrangement satisfactory, whereas others report that they are frustrated with their limited counseling and supervised time.

You may also want to consider whether a particular agency's schedules will mesh with your own needs. For example, some agencies offer group therapy sessions on several evenings during the week, which may conflict with your personal responsibilities.

Your Counseling Program Guidelines

As you begin exploring possible internships, be certain that your potential site fulfills all requirements of your counseling program. Most programs provide a clear, printed set of guidelines listing the expectations and regulations for student internships. For example, your counseling program guidelines may specify that interns acquire experience in treatment planning, in individual and group counseling, in case management, and in discharge planning, as well as gaining an understanding of agency administrative procedures. You should be sure that you will be able to satisfy all the necessary program requirements at each internship site. In addition, make certain to determine whether you are covered by malpractice insurance provided by your program. Your counseling program guidelines should specify whether or not students are covered. If you need to purchase malpractice insurance, professional organizations such as the American Counseling Association (ACA) provide coverage at reduced rates for students.

State Licensure/Certification Requirements

Most state counselor licensure/certification boards allow credit for supervised hours accrued during your internship, even before completion of your graduate program in counseling. However, there are specific rules and regulations concerning, for example, whether your experience may be paid or unpaid, the hours and nature of supervision, the relationship between supervisor and supervisee, and the intern's scope of practice. In addition, all state boards have formal procedures and policies for registering supervised counseling experience. As soon as you begin selecting an internship site, you should write or call your state counseling regulatory board to obtain all documents, applications, and instructions pertaining to counselor licensure/certification. Then read through everything carefully and consider whether you will be able to fulfill state requirements for supervised counseling experience at each internship site. In addition, once you have selected a site, make sure to follow through on all formal procedures to register your hours with your state board, so that you can get credit toward state licensure or certification for your internship.

Experiential and Didactic Opportunities of Each Site

Your internship is a critical part of the preparation for your career as a professional counselor, and you should carefully examine and analyze what kinds of educational opportunities you will be likely to experience at each potential site. Will you be having direct client contact, such as doing intake assessments and conducting individual or group counseling, or will much of your time be spent "running errands," such as filing papers and operating the copying machine? Will you be treated as a colleague and a valued member of the treatment team, or will the other professionals discount your input because you are "only a student"? Will you have the support you need, and will other staff members be willing to answer questions and offer help if you ask? Will you be invited to attend selected staff meetings and to participate in in-service educational sessions?

Your prospective supervisor will be responsible, in large part, for delineating your internship activities. Therefore, you should give thoughtful consideration to those personal and professional qualities you would hope to find in this individual. Your supervisor also must have the time, the interest, and the commitment to teaching interns, plus an understanding of the very special nature of the supervisory relationship.

SOURCES OF INFORMATION

As you begin the process of selecting your internship, you will need to assemble a list of potential sites. You may already have some thoughts about where you would like to do your internship. However, if you have no idea where to begin, we suggest a preliminary discussion with your counseling program coordinator of field placements. The coordinator should be able to provide helpful suggestions and insights concerning prospective placement sites and should be able to provide names of alumni of your program who are currently working at various sites and who would be good people to contact for further information. It may also be a good idea to sit down with your faculty advisor, who ideally knows something about your personality and needs, and who may be able to recommend several possible placements.

Networking with other students who are currently involved in a field placement or who have recently completed one is an invaluable way to gather information about potential sites. In this way, you may hear of several interesting sites that you simply did not know were available. You will probably also learn a great deal about a particular site's activities, schedule, staff, and atmosphere by talking to students who have worked there. Some of these same questions may be answered by site supervisors at the informational interview, which is discussed in the next section; however, it is often very helpful to have another student intern's perspective as well.

One additional source for potential field placement sites is the classified ad section of your local newspaper. The advertisements for help wanted, under such headings as "Social services," "Counselors," and "Mental health," often provide possible leads. You can telephone each agency and ask whether an internship or field placement program is available and the name of the person to contact for further information. In addition, many funding agencies, such as the United Way or local mental health board, publish listings of their affiliated agencies. Often these listings, which usually include agency service descriptions as well as contact persons, provide good leads for internships.

INFORMATIONAL INTERVIEWING

An awareness of your own needs, interests, and expectations may be acquired through course readings and experiences, as well as through actual visits to potential placement sites. Discussions with supervisors and clinicians at the site will enable you to clarify the four factors to consider in choosing your internship: (1) your own needs and interests, (2) your counseling program guidelines, (3) state regulations concerning counseling licensure/certification, and (4) the educational opportunities at each site. A visit to a potential site helps you determine whether that particular internship site is a good fit and will meet most of your expectations. This process, known as *informational interviewing,* can give you invaluable firsthand exposure to the unique environment of each of your choices.

To prepare for an informational interview, we suggest that you compose or update a résumé, outlining your educational and professional experience, including volunteer activities relevant to counseling or human services. (A sample résumé is provided in Appendix A.) Take along a copy of your résumé, plus any appropriate samples of your written work (such as case studies or research projects from courses), and also a copy of your counseling program's field experience regulations, including agency guidelines, course requirements for students, and evaluation procedures.

You should be prepared to answer, as well as to ask, questions when you go on your informational interviews; the professional at the placement site may look on this interview as an ideal opportunity to learn something about your personality and ability. Future supervisors have asked potential student interns a wide variety of questions, ranging from "What is your theoretical orientation?" to "What is your favorite restaurant?" to "Have you ever been in therapy yourself?" Your attitude toward being questioned and your style of relating to the interviewer may be more important than the actual answers to those questions.

Following the interview, we suggest that you send a brief thank-you note indicating your interest, if any, in the placement and noting that you will follow up with a phone call in one or two weeks to explore the possibility of arranging an internship at that site. (A sample thank-you note is provided in Appendix B.)

JOB DESCRIPTIONS

Once you have assembled a list of potential placement sites and have visited the most promising choices, you can begin comparing aspects of each placement to find one that best meets most of your needs and interests, fulfills your academic program specifications, satisfies as many of the state licensing/certification requirements as possible, and provides the optimal educational opportunities. To facilitate the selection process, you may find it helpful to prioritize criteria. For example, you may decide that it is more important to you to work with one specific client population than it is for you to work as a member of an interdisciplinary team. In addition, keep in mind that the goal is to find a good fit, rather than the perfect choice, so try to be flexible in assessing each placement site.

Job descriptions provide key information for you to consider, in conjunction with other data, because these represent the supervisor's or administrator's standards and expectations for each placement experience. You may be able to pick up actual printed job descriptions in the program office at your college or university, in the classified ad section of the newspaper, or in various agency offices. You may also find it helpful to take written notes on the verbal job descriptions given at informational interviews or during discussions with other students, advisors, or professors. These notes can be reviewed and compared later on. (A sample job description is provided in Appendix C.)

As you review these job descriptions and make your final selection for your internship, you may find it helpful to consider the following questions:

1. What kinds of people do I enjoy working with? Will I be likely to work with this population at this placement site?

2. What types of problems might I want to deal with? Will I encounter these types of problems here?
3. Is the experience paid or unpaid?
4. Does the placement allow for credit toward state licensure or certification requirements? Some state licensing/certification boards, for example, specify that the student must receive at least 1 hour of supervision for every 20 hours of client contact.
5. How much will I actually be working with clients, rather than observing or running errands?
6. Will I receive adequate individual supervision? Is the supervisor credentialed at the independent practice level of licensure or certification? Having met the supervisor, do I feel that we will be able to maintain a good working relationship?
7. Am I covered by liability insurance? Most institutions of higher education carry policies that insure students during their internships. In cases where the university or college does not provide insurance, students should seek coverage through other sources, such as professional organizations (for example, the ACA).
8. What is the general atmosphere at this placement site? Is it formal or informal? Will I be comfortable working here? Will I be welcomed as a colleague?
9. What other mental health professionals are on staff? Will I be directly involved with psychiatrists, psychologists, social workers, counselors, nurses, teachers, or child care workers? Will I be part of a treatment team?

GETTING READY FOR THE PLACEMENT

Finally! You think you have found an internship site that is a good fit for you, and your future supervisor has told you that you have the position. You wonder what to do next. First, you will need a written agreement for the placement, signed by your future supervisor, specifying the dates of your internship. Next, you should follow the prescribed state licensing/certification procedures to be sure to obtain credit for your supervised hours. This process usually involves having both you and your supervisor complete paperwork describing the proposed scope of practice, the setting, the number of hours of work and of direct supervision, and the supervisor's areas of expertise.

Your meeting with your supervisor, when these official forms are completed and signed, is a good time for you to ask, "What can I do, before I actually begin work, to best prepare myself for this placement?" Your supervisor may be able to offer some advice, for example, by suggesting a few books for you to read or by noting that you should thoroughly review the *Diagnostic and Statistical Manual* prior to beginning the placement experience. At this meeting with your supervisor, you may also want to discuss the actual scheduling of your work hours, so that an agreement acceptable to both of you can be worked out. Some students have found it useful to discuss dress codes, if applicable, and whether they will need to bring any special equipment or supplies. For example, one inpatient unit we know requires all counselors to use pink highlighters on their notes on the patients' charts to make these notes easy to see. At this internship site, interns come prepared with several pink highlighters each day.

You may also find it helpful, during the time before you begin your internship, to review academic coursework, concentrating on those areas you feel might require extra attention or study. The National Board of Certified Counselors (NBCC) provides coursework descriptions in the 10 areas of study relevant to counseling, and this may be used as a guideline for self-assessment (see Appendix D for the 1993 NBCC course descriptions). We also suggest a thorough reading of the American Counseling Association Ethical Guidelines for Counselors (see Appendix E), so that you will be ready to act in a professional manner in all situations.

A CHECKLIST FOR GETTING STARTED

To help you complete all the steps in the process of selecting your internship site, we have compiled the following checklist:

_____ 1. Determine your own interests, needs, and expectations for the internship.

_____ 2. Review your counseling or human services program regulations for internships.

_____ 3. Ascertain any state requirements for supervised work experiences related to counseling licensure/certification.

_____ 4. Think about what kinds of educational opportunities you hope to experience during your internship, and what kinds of qualities you would hope to find in your supervisor.

_____ 5. Make an appointment with your faculty advisor to discuss placement possibilities.

_____ 6. Make an appointment with the placement coordinator of your program to acquire further information.

_____ 7. Locate additional sources for information about potential internship sites, including other students, newspaper advertisements, and funding agencies.

_____ 8. Compose or update a résumé and make several typed copies.

_____ 9. Arrange informational interviews as often as possible.

_____ 10. Follow up on all interviews with a thank-you note and, a week or two later, a phone call.

_____ 11. Obtain and compare job descriptions for those internship sites that sound interesting to you.

_____ 12. Try to find a good fit for the placement, keeping in mind your own needs and interests, your program requirements, state licensing/certification regulations, and the unique experiences offered and the individual characteristics of each internship site.

_____ 13. Secure a final agreement with the supervisor after you have selected a placement site.

_____ 14. Complete all official forms to register your supervised experience with the state licensing/certification board.

_____ 15. Ask your future supervisor how you can best prepare for your placement. Follow through on his or her suggestions.

_____ 16. Carry out a self-assessment to discover academic areas relevant to the internship that you need to review.

_____ 17. Set up a schedule to read and review material, so that you will begin your placement feeling as competent and comfortable as possible.

Bibliography

BRADLEY, F. (Ed.). (1991). *Credentialing in counseling.* Alexandria, VA: American Counseling Association.

COLLISON, B., & GARFIELD, N. (1990). *Careers in counseling and development.* Alexandria, VA: American Counseling Association.

COREY, G. (1991). *Manual for theory and practice of counseling and psychotherapy.* Pacific Grove, CA: Brooks/Cole.

COREY, G. (1991). *Theory and practice of counseling and psychotherapy.* Pacific Grove, CA: Brooks/Cole.

GOLDBERG, C. (1992). *The seasoned psychotherapist: Triumph over adversity.* New York: W. W. Norton.

HEPPNER, P. (Ed.). (1990). *Pioneers in counseling and development: Personal and professional perspectives.* Alexandria, VA: American Counseling Association.

HERR, E. (1989). *Counseling in a dynamic society: Opportunities and challenges.* Alexandria, VA: American Counseling Association.

JOHNSON, M., CAMPBELL, J., & MASTERS, M. (1992). Relationship between family of origin dynamics and a psychologist's theoretical orientation. *Professional Psychology: Research and Practice, 23*(2), 119–122.

LEWIS, M., HAYES, R., & LEWIS, J. (1986). *The counseling profession.* Itasca, IL: Peacock.

MYERS, W. (1992). *Shrink dreams.* New York: Simon and Schuster.

PECK, M. (1978). *The road less traveled.* New York: Simon and Schuster.

SIEGEL, S., & LOWE, E. (1992). *The patient who cured his therapist.* New York: Dutton.

SKOVHOLT, T., & RONNESTAD, N. (1992). Themes in therapist and counselor development. *Journal of Counseling and Development, 70*(4), 505–515.

SUSSMAN, M. (1992). *A curious calling: Unconscious motivation for practicing psychotherapy.* Northvale, NJ: Jason Aronson.

WALLACE, S., & LEWIS, M. (1990). *Becoming a professional counselor.* London: Sage.

YALOM, I. (1974). *Every day gets a little closer.* New York: Basic Books.

YALOM, I. (1989). *Love's executioner.* New York: Basic Books.

CHAPTER TWO

DEVELOPING COMPETENCIES AND DEMONSTRATING SKILLS

Your internship provides an arena for you to try your wings as a helping professional, with guidance and support close at hand. Many students feel a bit overwhelmed as they begin to interact with clients and staff members. You may have difficulty at first, as you try to remember counseling theory and techniques, to recall academic coursework concerning such areas as human development or multicultural issues, to keep in mind ethical guidelines, and to think about agency procedures, regulations, and policies—all while trying to attend to your first few clients! More than one intern has felt discouraged after the first week of trying to juggle all the responsibilities of the new role.

Your internship can be viewed as a time to build a framework of new professional relational skills on a foundation of the material you have learned in your counseling program courses, your own life experiences, and your personal values and philosophies. This framework is composed of new perspectives, understandings, abilities, and skills, added gradually and with care. Your goal is to construct a strong framework over a solid foundation, working diligently but patiently, and often standing back to take a look at the work you have accomplished so far.

During your internship, you will be developing some of the specific personal attributes and professional competencies that you will use during your professional counseling career. To help you delineate your goals, we have compiled the following list of skills for graduate-level internship students to work toward building. In reality, not all placement sites afford the opportunity to develop abilities in every area we have indicated. In addition, the quantity and scope of the competencies listed below reflect our implicit belief that the process of becoming a professional counselor is an ongoing one. Your internship is just the beginning of your professional development; you will continue to add competencies throughout your career.

SUGGESTED COMPETENCIES FOR INTERNS

I. Communication Skills
  A. Verbal skills
    1. Students will be able to express themselves clearly and concisely in daily interactions with agency staff members and other professionals.
    2. Students will be able to communicate pertinent information about clients and to participate effectively in interdisciplinary treatment team meetings and case conferences, while maintaining their identities as counselors within a multidisciplinary group.
    3. Students will be able to educate clients and to provide appropriate information on a variety of issues (such as parenting, after-care and other support services, psychotropic medications, stress management, sexuality, psychiatric disorders) in an easily understandable manner.
    4. Students will be able to communicate with clients' families, significant others, and designated friends in a helpful fashion. They will be able to provide, as well as to obtain, information concerning the client, while respecting the client's rights concerning privacy, confidentiality, and informed consent.
    5. Students will be able to communicate effectively with referral sources, both inside and outside the agency, concerning all aspects of client needs and well-being (for example, housing, legal issues, Twelve Step programs, psychiatric concerns).
  B. Writing skills
    1. Students will be able to prepare a complete, written initial intake assessment, including a mental status evaluation, a psychosocial history, a diagnostic impression, and recommended treatment modalities.
    2. Students will be able to write progress notes, to chart, and to maintain client records according to agency standards and regulations.
    3. Students will be able to prepare a written treatment plan, including client problems, therapeutic goals, and specific interventions to be utilized.
    4. Students will be able to prepare a formal, written case study.
    5. Students will be able to use computer skills to work with word-processing programs and to maintain and search data bases.
  C. Knowledge of nomenclature
    1. The student will acquire a thorough knowledge of professional terminology pertaining to counseling, psychopathology, treatment modalities, and psychotropic medication.

2. The student will be able to understand professional counseling jargon and will be able to participate in a professional dialogue.

II. Interviewing

A. The student will structure the interview according to a specific theoretical perspective (for example, psychodynamic or behavioral theory), because a theory base provides the framework and rationale for all therapeutic strategies, techniques, and interventions.

B. The student will be able to use appropriate counseling techniques to engage the client in the interviewing process, to build and maintain rapport, and to begin to establish a therapeutic alliance. This may include using attending behaviors, active listening skills, and a knowledgeable and professional attitude to convey empathy, genuineness, respect, and caring, and to be perceived as trustworthy, competent, helpful, and expert (Lewis, Hayes, & Lewis, 1986).

C. The student will be able to use appropriate counseling techniques to increase client comfort and to facilitate collection of data necessary for clinical assessment, such as conducting a mental status evaluation, taking a thorough psychosocial history, and eliciting relevant, valid information concerning the presenting problem, in order to formulate a diagnostic impression. Specific interviewing competencies may include observation, use of open-ended and closed-ended questions, the ability to help the client stay focused, reflection of content and feeling, reassuring and supportive interventions, and the ability to convey an accepting and nonjudgmental attitude.

D. The student will develop a holistic approach toward interviewing by assessing psychological and biological factors, as well as environmental and interpersonal factors, that may have contributed to the client's developmental history and presenting problems.

E. The student will strive to see things from the client's frame of reference and to develop a growing understanding of the client's phenomenological perspective.

F. The student will be aware at all times of the crucial importance of understanding the client from a multicultural perspective and will be aware that sociocultural heritage is a key factor in determining the client's unique sense of self, worldview, values, ideals, patterns of interpersonal communication, family structure, behavioral norms, and concepts of wellness as well as of pathology.

III. Diagnosis

A. The student will acquire an understanding of the most commonly used assessment instruments, such as personality and intelligence tests, anxiety and depression scales, and interest inventories.

1. The student will acquire a familiarity with the validity and reliability of these instruments.
2. The student will be able to interpret data generated by these instruments and to understand the significance of these data in relation to diagnosis and treatment.
3. The student will be able to determine which assessment instruments would be most helpful in evaluating specific client problems or concerns.
4. The student will be aware of the limitations of assessment instruments when used with ethnic minority populations.

B. The student will develop a working knowledge of the DSM-IV.

1. The student will be familiar with the organization of the DSM-IV and will be able to use this nosology effectively (for example, to find diagnostic codes or to trace clients' behaviors, affects, or cognitions along the decision trees to ascertain potential diagnoses).
2. The student will be able to understand the DSM-IV classification of disorders and will be able to identify particular constellations of client problems as specific DSM-IV diagnostic categories.

C. The student will be able to review and consider all pertinent data, including interviews, medical records, previous psychiatric records, test results, psychosocial history, consultations, and DSM-IV classifications, in formulating a diagnostic impression or preliminary diagnosis.

IV. Treatment

A. Students will be able to conduct therapy using accepted and appropriate treatment modalities and counseling techniques based on recognized theoretical orientations.

1. Students will work toward identifying their own theoretical framework based on their own philosophy of humankind.
2. Students will know how to make treatment recommendations, formulate a treatment plan, establish a treatment contract, implement therapy, and terminate the therapeutic relationship at an appropriate time.
3. Students will be able to conduct the following types of therapy and will understand the underlying principles, issues, dynamics, and role of the counselor associated with each type of treatment:
   a. individual therapy
   b. marital therapy
   c. conjoint therapy
   d. family therapy

       e. group therapy

       f. crisis intervention

  B. Students will understand that different client populations and different types of problems may best respond to varying therapeutic approaches and techniques.

     1. Students will be knowledgeable about various types of client populations and their particular problems and concerns, including but not limited to the following:

       a. children and adolescents

       b. adults

       c. the elderly

       d. chemically dependent individuals

       e. adult children of alcoholics

       f. gay and lesbian clients

       g. survivors of trauma and abuse

       h. eating disordered individuals

       i. physically or cognitively impaired clients

       j. dual-diagnosed clients (for example, chemically dependent with a psychiatric disorder)

       k. clients of varied ethnic backgrounds

     2. Students will be able to be flexible and knowledgeable in determining population-appropriate counseling techniques and therapeutic interventions. Students will have as many therapeutic tools available for use as possible (for example, play therapy, art therapy, behavioral techniques, role playing, gestalt techniques, directive versus nondirective techniques, stress-management techniques, experiential therapy, hypnosis).

  C. Students will be sensitive to the impact of multicultural issues on the counseling relationship and on treatment and will modify therapeutic approaches and techniques to respect multicultural differences and to meet multicultural needs.

V. Case Management

  A. The student will acquire an understanding of the functions and goals of all departments, programs, and services within the agency and will be able to network with appropriate personnel throughout the social service system.

  B. The student will understand the roles, responsibilities, and contributions to client care of members in each department or program within the agency. The student will know which individual(s) to contact to help resolve various client problems.

  C. The student will acquire a thorough knowledge of community resources and will understand the agency procedures for referring clients to outside sources for help.

  D. The student will consider continuity of care to be a most important goal, beginning with the initial client contact.

1. The student will act as an advocate for the client in ensuring continued quality of care and access to social services. Advocacy will include, but not be limited to, exploring possible funding sources for care, such as mental health coverage on insurance policies or Medicaid or Medicare.
2. The student will be able to participate in all areas of discharge planning, including arranging follow-up visits with a mental health professional, communicating with insurance companies, and providing help with housing, transportation, vocational guidance, legal assistance, support groups, medical care, and referral to other services or agencies.

VI. Agency Operations and Administration
A. The student will be familiar with the organizational structure of the agency and will understand the responsibilities and functions of administrative staff.
B. The student will understand the philosophy, mission, and goals of the agency and will have a thorough knowledge of all policies and procedures of the agency, which are usually delineated in a comprehensive manual.
C. The student should be aware of immediate and long-range strategic plans for the agency (for example, to hire an art therapist, to develop a chemical abuse program, or to add an additional building, as well as to evaluate and eliminate ineffective programs).
D. The student will have an understanding of the business aspects of the agency (for example, funding sources or budget allowances).
E. The student will be aware of legal issues concerning agency functions, such as state or national licensure/certification requirements or safety regulations.
F. The student will understand agency standards to ensure continued quality of care, including quality assurance and peer review processes.

VII. Professional Orientation
A. The student will be knowledgeable concerning all ethical and legal codes for counselors, provided by professional counseling associations as well as by state law, and will adhere to these standards at all times.
B. The student will be familiar with agency regulations and policies regarding ethical and legal issues and will adhere to these standards at the placement site.
C. The student will be knowledgeable concerning legislation protecting human rights.
D. The student will seek guidance from the on-site supervisor and the academic program supervisor with any questions concerning ethical or legal issues or professional behavior.

E. The student will consider the four basic Rs for counselors (Carkhuff, cited in Lewis, Hayes, and Lewis, 1986) whenever acting in a professional helping capacity: the *right* of the counselor to intervene in the client's life, the *responsibility* the counselor assumes when intervening, the special *role* the counselor plays in the helping process, and the *realization* of the counselor's own resources in being helpful to the client.

The reader may wish to refer to a sample intern evaluation form in Appendix H.

References

LEWIS, M., HAYES, R., & LEWIS, J. (1986). *The counseling profession.* Itasca, IL: Peacock.

Bibliography

AMERICAN PSYCHIATRIC ASSOCIATION. (in press). *Diagnostic and statistical manual of mental disorders* (4th ed.). Washington, DC: Author.

ANDERSON, D. (1992). A case for standards of counseling practice. *Journal of Counseling and Development, 71*(1), 22–26.

AXLINE, V. (1969). *Play therapy.* New York: Random House.

BALSAM, M., & BALSAM, A. (1984). *Becoming a psychotherapist* (2nd ed.). Chicago: University of Chicago Press.

BENJAMIN, A. (1987). *The helping interview.* Boston: Houghton Mifflin.

BURN, D. (1992). Ethical implications in cross-cultural counseling and training. *Journal of Counseling and Development, 70*(5), 578–583.

CARKHUFF, R. (1987). *The art of helping* (Vol. VI). Amherst, MA: Human Resource Development Press.

CASEMENT, P. (1991). *Learning from the patient.* New York: Guilford.

CHAPLIN, J. (1985). *Dictionary of psychology* (2nd ed. rev.). New York: Doubleday.

CHESSICK, R. (1989). *The technique and practice of listening in intensive psychotherapy.* New York: Jason Aronson.

CHESSICK, R. (1993). *A dictionary for psychotherapists: Dynamic concepts in psychotherapy.* Norwalk, NJ: Jason Aronson.

ELKIND, S. (1992). *Resolving impasses in therapeutic relationships.* New York: Guilford.

ENGELS, D., & DAMERON, J. (Eds.). (1990). *The professional counselor: Competencies, performance, guidelines, and assessment* (2nd ed.). Alexandria, VA: American Counseling Association.

FAIVER, C. (1992). Intake as process. *Journal of the California Association for Counseling and Development, 12,* 83–86.

FALVEY, J. (1992). From intake to intervention: Interdisciplinary perspectives on mental health treatment planning. *Journal of Mental Health Counseling, 14*(4), 471–489.

HOOD, A., & JOHNSON, R. (1991). *Assessment in counseling: A guide to the use of psychological assessment procedures.* Alexandria, VA: American Counseling Association.

HUGHES, J., & BAKER, D. (1990). *The clinical child interview.* New York: Guilford.

KARASU, T. (1992). *Wisdom in the practice of psychotherapy.* New York: Basic Books.

KOTTLER, J. (1992). *Compassionate therapy: Working with difficult clients.* San Francisco: Jossey-Bass.

KURPIUS, D., & BROWN, D. (1988). *Handbook of consultation: An intervention for advocacy and outreach.* Alexandria, VA: American Counseling Association.

L'ABATE, L., FARRAR, J., & SERRITELLA, D. (Eds.). (1992). *Handbook of differential treatments for addictions.* Boston: Allyn and Bacon.

LEE, C., & RICHARDSON, B. (1991). *Multicultural issues in counseling.* Alexandria, VA: American Counseling Association.

MITCHELL, R. (1991). *Documentation in counseling records* (Vol. 2). Alexandria, VA: American Counseling Association.

MORROW, K., & DEIDAM, C. (1992). Bias in the counseling profession: How to recognize and avoid it. *Journal of Counseling and Development, 70*(5), 571–577.

OGDON, D. (1990). *Psychodiagnostics and personality assessment: A handbook* (2nd ed.). Los Angeles: Western Psychological Services.

PARAD, H., & PARAD, L. (1990). *Crisis intervention* (Vol. 2). Milwaukee: Family Service America.

PEDERSON, P. (1988). *A handbook for developing multicultural awareness.* Alexandria, VA: American Counseling Association.

REID, W., & WISE, M. (1989). *DSM-III-R Training Guide.* New York: Brunner/Mazel.

SHAPIRO, D. (1965). *Neurotic styles.* New York: Basic Books.

SHEA, S. (1988). *Psychiatric interviewing: The art of understanding.* Philadelphia: Saunders.

SPITZER, R., GIBBON, M., SKODOL, A., WILLIAMS, J., & FIRST, M. (1989). *DSM-III-R Case Book.* Washington, DC: American Psychiatric Press.

STEENBARGER, B. (1991). All the world is not a stage: Emerging contextualist themes in counseling and development. *Journal of Counseling and Development, 70*(2), 288–296.

STREAN, H. (1993). *Resolving counterresistances in psychotherapy.* New York: Brunner/Mazel.

SWARTZ, S. (1992). Sources of misunderstandings in interviews with psychiatric patients. *Professional Psychology: Research and Practice, 23*(1), 24–29.

WALLACE, S., & LEWIS, M. (1990). *Becoming a professional counselor.* London: Sage.

YALOM, I. (1985). *Theory and practice of group psychotherapy.* New York: Basic Books.

CHAPTER THREE

THE SITE SUPERVISOR

GENERAL RESPONSIBILITIES

The internship site supervisor serves in a highly significant role and has many responsibilities for your training. With this individual, you will begin to demonstrate the knowledge and skills you have acquired through your formal academic training. The professional and personal relationship you establish with your site supervisor will set the pace, direction, and tone for your internship. His or her willingness to supervise you becomes a commitment to assist you in attaining and maintaining a counseling relationship with clients.

Ronnestad and Skovholt (1993) noted that beginning students value a supervisor who teaches and provides structure and direction to them. Other researchers emphasize the importance of supervisor support of interns, which may include such items as assistance with the selection of clients and provision of relevant agency and client information (a guidance function) (Grater, 1985); establishment of a milieu that creates a sense of security for the intern (Rabinowitz, Heppner, & Roehlke, 1986); and the provision of encouragement and feedback as to progress (Worthington & Roehlke, 1979).

In many ways what follows is our conception of the ideal supervisor, the super supervisor. Let's call this person the SUPERvisor. This is certainly unfair to supervisors in that it is not always possible to demonstrate perfection in deed or in person. Nonetheless, for our purposes, we shall demand perfection!

You must feel professionally and personally comfortable with your SUPERvisor and believe that this person will be a good role model for you. The time you spend together should provide ample opportunities for you to get to know each other as well as to assess your ability to work together. Even though the SUPERvisor maintains major responsibility for the integrity of the internship and for encouraging you as you pursue a career in counseling, your belief that he or she can work with you is vital. Ideally, the SUPERvisor and you share the hope that the internship will be a worthwhile and enjoyable experience for both of you. Any hesitancy you have should be discussed candidly with your site SUPERvisor or university SUPERvisor so any necessary accommodations can be made.

The internship experience in many ways is similar to the counseling relationship; as such, you will want to work with a SUPERvisor whose therapeutic orientation, ideology, and style are similar to yours.

One of the key roles your SUPERvisor will perform is to provide feedback as to your performance and progress. This critical feedback process will help you stay focused on both the quality and the quantity of your services. As part of the ACA Code of Ethics (see Appendix E, section H, number 4), your SUPERvisor accepts the responsibility to help ensure that you develop into a competent counselor. By providing feedback and direction he or she provides a great preparatory service in your professional development.

As a SUPERvisor, he or she is an educator and an extension of your formal academic training. The site SUPERvisor's role is unique in that his or her role modeling may influence your approach to the profession. Our ideal SUPERvisor assists you in feeling comfortable in awkward or new situations. Moreover, we hope that you experience your SUPERvisor's humanity as well as his or her professional persona and that he or she allows you to witness an individual who feels comfortable in letting you know the intricacies of the profession by his or her honest display of attitude, emotions, and behavior. You will find his or her support and direction key factors in the refinement of your continuing journey into the counseling profession. The site SUPERvisor's task is of a serious nature; he or she has made a commitment to counseling itself by agreeing to supervise you.

At times, you may find that the site SUPERvisor will share facets of the profession that you have not learned in textbooks. We anticipate that his or her experience, training, and skills will afford you the opportunity to learn much about the field. In essence, he or she might share "the good, the bad, and the ugly" of the counseling field, all of which can be valuable as you continue your professional development. Take all experiences as learning experiences; even the bad can be looked on (as some say, reframed) as a challenge and opportunity.

An issue students often have to address is how to interact with other mental health professionals at the site. Most likely, your SUPERvisor will ask his or her colleagues to expose you to their professional style, approaches, and theories. Through such exposure, you may find that, at times, their vantage point will differ from your SUPERvisor's. This can pose a dilemma and, perhaps, be confusing to you. Our ideal SUPERvisor is acutely aware that his or her orientation is not absolute. However, he or she will likely desire that you adhere to his or her directives, because ultimate responsibility rests with the SUPERvisor. He or she will be more than willing to discuss with you any differences you have observed and, generally, will use such experiences as teaching tools. Ultimately, by internal (sometimes unconscious) reflection you will decide where all these bits of information, observations, and various approaches fit you as a therapist. In the interim, you will save yourself much frustration and cognitive dissonance by relying on this SUPERvisor to achieve your educational goals and objectives. Work closely with your SUPERvisor to make the internship experience an enjoyable, educational, and personally rewarding opportunity. Do not hesitate to share with him or her where your interests lie and what you like and dislike about the total experience, including agency ambience, clients, and staff. Being candid with your SUPERvisor can make this experience an optimal one, enhancing both your learning and your relationship. Your placement serves as a "testing ground" for academic knowledge and should be as pleasant and beneficial as possible for the best possible learning results. No doubt your SUPERvisor will share this goal with you.

As noted previously, this is probably the final stage of your academic endeavor and, as such, is the pinnacle of a course of study. Among the many thoughts that might occur to you at this point is, "Do I really want to be a counselor?" Don't be concerned about such thoughts. It is our experience that this may be a reflection of self-doubt. Be willing to explore thoughts such as these with your SUPERvisor. Also, integrating your formal education with practice may cause you to realize that the counseling field may not be exactly what you thought, were told, or read it would be. Believing something about the profession without validating it does a great disservice to yourself, your colleagues, and, above all, future clients. Use the placement as the arena for solidifying your beliefs and attitudes about the profession. Because your SUPERvisor and university academic advisor have worked with you up to this point to establish your professional goals and objectives, it is most appropriate to discuss this key factor as well.

Oftentimes, separating your professional and personal involvement with your SUPERvisor is difficult, because the nature of the placement can cause blurred role boundaries. Your SUPERvisor adheres closely to the ACA Code of Ethics (see Appendix E), so his or her personal side can influence the tone, atmosphere, and ethos of the placement. Exposing his or her shortcomings as well as strengths can only serve to reinforce the uniqueness and value of your relationship. Further, being a person first enhances the value of the quality of interaction.

It is a rare SUPERvisor who does not learn from his or her students. For example, your current involvement in the academic arena may serve to augment his or her knowledge base. Your SUPERvisor may take pride in helping to shape and mold your career, certainly a powerful personal learning experience. The relationship you share can leave a positive and lasting impression on you both.

PRIOR TO INTERNSHIP

Now, let's look at this special relationship as it develops from beginning to end. A number of tasks need to be achieved when you first meet with your SUPERvisor. He or she will need to know your university and state licensing or certification requirements as well as your personal goals. The placement can influence your belief about the counseling profession itself. Thus, it may be helpful for the SUPERvisor to require definitive, objective, and measurable goals. Additionally, he or she has the responsibility to gain an understanding of who you are, especially as an aspiring counseling professional. Because the SUPERvisor is aware of the placement site's policies and procedures, he or she will gauge the appropriateness of a match and the potential for successful completion. It is not unusual for the SUPERvisor to conduct a somewhat formal interview with you to gain an understanding of why you chose this site, what your career objectives are, and what you would like to achieve in your professional relationship with him or her. Your level of understanding regarding theories of counseling, techniques, and the counseling field in general will also be of importance. The SUPERvisor will assess whether the melding of book knowledge from your academic training can occur at this specific site. In addition to examining your academic foundation, your SUPERvisor will also take into consideration your personal attributes.

No two training sites have the same mix of personnel, client population, and agency standards and ethos. For these reasons, the SUPERvisor, as presumed expert of his or her professional domain, must make the final disposition regarding your acceptance.

Once you are accepted, however, our ideal SUPERvisor will outline for you any forms you need to complete prior to starting the placement. The SUPERvisor may also suggest specific readings, including pertinent agency orientation materials, such as policies and procedures. Perhaps you will be asked to review your theories and techniques as well as to start solidifying a therapeutic orientation based on your knowledge of the clinical population. You may also be asked to prepare audiotape or videotape role-play demonstrations so the SUPERvisor can establish a baseline of your knowledge and skill level. The SUPERvisor may also want to discuss with you how you feel personally about working with certain clinical populations to examine if there are any personal issues that may impede your ability to be effective and efficient with clients. Professional involvement will not be void of the personal feelings, attitudes, and beliefs that are a part of your individuality—part of who you are.

As the final preparations are made for your placement, the SUPERvisor will explain in depth the agency's policies, procedures, mission statements, and other information relevant to your placement. Specific attention may be given to past involvement of students at this site and how they fit into the overall functioning of the organization. The SUPERvisor helps you to understand how you will be perceived during your placement and what specific parameters will be in place, how you will "fit" into the organization. Your SUPERvisor will be an advocate for you and has agreed to shoulder the responsibility to have you integrate successfully into the treatment milieu and the organizational structure. The routine and seemingly mundane tasks of completing the necessary academic course forms, including registration, and student/SUPERvisor agreements should be done at your earliest convenience. Watch for university and agency deadlines! For those individuals governed by specific licensure and/or certification requirements, understanding these standards, including the prompt filling out of any necessary forms, is vital to receive proper experiential credit.

DURING INTERNSHIP

Because internship is a great opportunity to put your skills and knowledge to use, our exemplary SUPERvisor will provide you with the chance to use them to the fullest. The ideal is to have you experience all the clinical aspects possible within the organization. What better way to delve into the "real world" of counseling than to practice one's skills in all areas of clinical services.

Barring any sensitive issues for which clients may prefer not to have you present, it is appropriate and a valuable learning experience to "shadow" the SUPERvisor in all of his or her clinical activities. By observing the SUPERvisor, including the eventual participation in services, you will gradually become acclimated to the setting and begin integrating your book knowledge with clinical practice. Shadowing provides both of you the opportunity to assess the scope of your knowledge and level of comfort as you aspire to independent work with clients. When you accept the responsibility of independent client contact depends on a consensus between you and your SUPERvisor as to your readiness. Further, unfortunately, our clients do not always fit into our plans as to when and what types of techniques or approaches we would like to "practice" with them. Feel comfortable to discuss with your SUPERvisor your perceived readiness to accept more responsibility and to use your knowledge and skills. Take note of how it feels to put your knowledge to use.

One critical issue to address throughout the internship is what treatment approach to use with your clients. Before you solidify your approach, review with your SUPERvisor your "philosophy of humankind" (see chapter 9 on self-assessment). Perhaps you have previously dealt with this personal philosophical stance in your theoretical studies, but now you link those theories with your belief system in a practical setting with contact with fellow human beings, who are, for the most part, suffering. We believe that your philosophy of humankind has an effect on the way you approach and treat your clients, and, for that matter, people in general. For many students this will be the first time they closely examine how their beliefs can influence their dealings with clients in vivo. Once you have begun to formulate, review, and refine your philosophy of humankind, our ideal SUPERvisor can guide you in the selection of a philosophically and theoretically congruent treatment approach. For example, existential approaches lend themselves more to freedom of choice, whereas analytically and behaviorally oriented approaches appear to be deterministic.

Throughout the internship your SUPERvisor will be scheduling both formal and informal supervision conferences with you. During these meetings you will be reviewing your internship goals and objectives to assess adherence, progress, and any necessary modifications to them. He or she will provide you with constructive feedback as to the quality of your performance, including recommendations for improvement. You will generally be formally evaluated midway through the placement and again at the end. Your academic advisor may be part of this process as well. Particular attention may be given to how well you feel you become part of the agency treatment structure and the overall treatment team. Your SUPERvisor will validate your perceptions of these two areas. In addition, your SUPERvisor may help you consider whether a specific clinical population or treatment specialty may be indicated and elaborate on any areas in which you are demonstrating competency (see chapter 2, Developing Competencies and Demonstrating Skills).

AT THE TERMINATION OF INTERNSHIP

Your SUPERvisor has served in a highly significant role and maintained a considerable amount of responsibility throughout your internship. One of the most enjoyable tasks he or she has to perform is to assist you in bringing closure to your placement experience. Your SUPERvisor has worked closely with you by investing time, knowledge, and skill to help you become a competent counselor in training. He or she will be eager to share impressions and recommendations with you as to your clinical competencies, skills, and knowledge base. The SUPERvisor will provide feedback as to how well you were able to translate your academic foundation to the practical work situation. His or her input, guidance, and direction can have an influence on your career focus. This is a responsibility in which the SUPERvisor takes pride. The completion of your final evaluation, often in concert with your academic advisor, will serve as the formal appraisal of your efforts. This formal document presents measurable, objective, and observable data as to your clinical strengths, weaknesses, general impressions, and overall recommendations. You should have ample opportunity to respond to the evaluation and, if necessary, get clarification on its content.

References

GRATER, H. (1985). Stages in psychotherapy supervision: From therapy skills to skilled therapist. *Professional Psychology: Research and Practice, 16,* 605–610.

RABINOWITZ, F., HEPPNER, P., & ROEHLKE, H. (1986). Descriptive study of process and outcome variables of supervision outcome. *Journal of Counseling Psychology, 33*(3), 292–300.

RONNESTAD, M., & SKOVHOLT, T. (1993). Supervision of beginning and advanced graduate students of counseling and psychotherapy. *Journal of Counseling and Development, 71*(4), 396–405.

WORTHINGTON, E., & ROEHLKE, H. (1979). Effective supervision as perceived by beginning counselors in training. *Journal of Counseling and Development, 26*(1), 64–73.

Bibliography

ALLEN, G., SZOLLOS, S., & WILLIAMS, B. (1986). Doctoral students' comparative evaluations of best and worst psychotherapy supervision. *Professional Psychology: Research and Practice, 17,* 91–99.

BARTLETT, W., GOODYEAR, R., & BRADLEY, F. (Eds.). (1983). Supervision in counseling II (Special issue). *The Counseling Psychologist, 11*(1).

BORDERS, L., & LEDDICK, G. (1987). *Handbook of counseling supervision.* Alexandria, VA: American Counseling Association.

ELLIS, M. (1991). Critical incidents in clinical supervision and in supervisor supervision: Assessing supervisory issues. *Journal of Counseling Psychology, 38*(3), 342–349.

GLIDDEN, C., & TRACEY, T. (1992). A multidimensional scaling analysis of supervisory dimensions and their perceived relevance across trainee experience levels. *Professional Psychology: Research and Practice, 23*(2), 151–157.

HESS, A. (1980). *Psychotherapy supervision: Theory, research and practice.* New York: Wiley.

LEWIS, M., HAYES, R., & LEWIS, J. (1986). *An introduction to the counseling profession.* Itasca, IL: Peacock.

MASTERS, M. (1992). The use of positive reframing in the context of supervision. *Journal of Counseling and Development, 70*(3), 387–390.

MUNSON, C. (Ed.). (1984). *Supervising student internships in human services.* Binghamton, NY: Haworth.

PITTS, J. (1992). Pips: A problem-solving model for practicum and internship. *Counselor Education and Supervision, 32*(2), 142–150.

SKOVHOLT, T., & RONNESTAD, M. (1992). Themes in therapist and counselor development. *Journal of Counseling and Development, 70*(4), 505–515.

SWANSON, J., & O'SABEN, C. (1993). Differences in supervisory needs and expectations by trainee experience, cognitive style, and program membership. *Journal of Counseling and Development, 71*(4), 457–464.

VASQUEZ, M. (1992). Psychologist as clinical supervisor: Promoting ethical practice. *Professional Psychology: Research and Practice, 23*(3), 196–202.

YERUSHALMI, H. (1992). On the concealment of the interpersonal therapeutic reality in the course of supervision. *Psychotherapy, 29*(3), 438–446.

CHAPTER FOUR

DECIDING HOW TO HELP: CHOOSING TREATMENT MODALITIES

By the time you are ready to begin your internship, you will have completed all or most of the academic courses required for your degree in counseling. You will have studied a variety of theoretical models and clinical techniques, and you will also have acquired a substantial data base in many other scholastic areas that are fundamental to counseling. This chapter provides a practical framework to help you synthesize all this academic information and formulate a rational protocol for selecting treatment modalities. In addition, we will highlight the theoretical concepts and specific counseling methods we feel have the most salience for you as a counselor intern.

THE COUNSELING PROCESS

Professional counseling may be conceptualized as an interactive process between counselor and client that enhances the client's level of functioning. The helping process "consists of a partnership of counselor and client working together to solve client problems by activating client assets, developing client skills, and utilizing environmental resources in order to decrease client problems and increase client coping skills" (Hershenon & Power, cited in Wallace & Lewis, 1990, p. 57). However, when you are faced with a client who presents a myriad of problematic behaviors, seems to have no social support system, and says, "I'm a complete failure; there's nothing I like about myself," you may wonder just where and how to begin helping.

When working with clients during your internship, remember that the counseling process always involves the following six stages, regardless of your theoretical orientation, your level of clinical experience, or the complexity of the client's problems:

1. Developing trust
2. Exploring problem areas
3. Helping to set goals
4. Empowering into action

5. Helping to maintain change
6. Agreeing when to end (Velleman, 1992)*

These stages are not necessarily linear or sequential at all times. For example, difficulties at Stage 4 (empowering into action) may lead to further work at Stage 2 (exploring problem areas) or at Stage 3 (setting goals).

The development of trust, the first stage, is the key factor in the counseling process and in fact is the underlying force that enables the counselor and client to negotiate the other stages (Velleman, 1992). Trust may need to be tested by the client and reestablished by the counselor many times, depending on the client's personality, developmental history, and current life situation. Often, each time the client faces a challenging new issue or a painful or frightening feeling, trust is questioned once again. When you select a treatment modality, you will need to consider the current level of trust your client feels; some therapeutic interventions may be too uncomfortable for clients who are feeling unsafe.

ESTABLISHING TRUST

As an intern, you can begin counseling by working to help your client feel safe. Always be reliable and dependable. Prove your trustworthiness in basic, concrete ways. We have found the following behaviors to be especially helpful in developing client trust:

- Start and end sessions on time, and do your best to maintain the regularity of the appointment schedule that you and your client have agreed upon.
- Make certain that your client has your full attention and respect during sessions, and try to demonstrate your caring in both verbal and nonverbal ways.
- Always follow through on counseling-related tasks, such as finding a support group or a list of helpful books for your client.
- Always maintain strict confidentiality; never discuss a client outside of the treatment setting unless you first have a signed release and have explained to your client exactly what information will be disclosed or exchanged.
- Do not talk about the problems of other clients, even without mentioning their names, during counseling sessions. At the least, many clients find hearing about other clients intrusive; others find it very upsetting indeed.
- Maintain professional boundaries at all times: do not touch your client physically; do not engage in any social interactions outside of the counseling relationship; do not disclose intimate personal information

*From *Counselling for Alcohol Problems,* by R. Velleman. Copyright © 1992 by Sage Publications, Inc. Reprinted by permission.

or discuss your own problems; do not participate in any business transactions with your client except for specified agency arrangements for the client to pay for services (Gutheil & Gabbard, 1993).

THERAPEUTIC TOOLS

During your internship, you will be working to develop a repertoire of therapeutic tools to help your client through each of the counseling stages mentioned above. These tools include your personal style of engaging a client, your ability to attend to both verbal and nonverbal cues in order to understand the client fully, and your specific counseling methods and interventions. The tools you choose should feel comfortable to you and should be appropriate for the particular client, the client's unique concerns, and the current stage of the counseling process.

A wide range of therapeutic tools and flexibility in selecting them help ensure that you will have plentiful resources to assist your clients during your internship. We are not advocating a random assortment of counseling techniques or an undisciplined eclecticism with no regard for the principles of the theoretical schools. We suggest, instead, that your choice of techniques be *grounded* in your knowledge of counseling theory, while it is *guided* by

- the individual needs and goals of the client
- the immediate problems of the client
- the present stage of the counseling process
- your evolving relationship with the client
- your personal philosophy of life and understanding of humankind.

Corey (1991a) recommended an integrative perspective for the student counselor: "Comparing your own thinking on the underlying issues with the positions of the various therapies can assist you in developing a frame of reference for your personal style of counseling" (p. 178).

Your personal counseling style is in part determined by your activity level in counseling, which includes the frequency of your interventions and your role in controlling the direction of the session/therapeutic process. The various therapies may also be arranged along a continuum of counselor activity level, with classic psychoanalytic schools on the less active end and strict behavioral schools on the more active end (see Figure 4.1). You may wish to compare your own comfort level and personal style with the counselor activity levels of the theoretical schools.

THE THERAPEUTIC RELATIONSHIP

Virginia Satir (1987) wrote that, although many diverse theories and techniques of therapy have been developed, "the basic ingredient remains the relationship between the therapist and the patient" (p. 17). The importance of the helping relationship itself as a healing factor and as a pathway for client change,

More Counselor Activity

Behavioral therapies

Cognitive/behavioral therapies

Gestalt therapy

Existential therapies

Client-centered therapy

Psychodynamic therapies

Classic psychoanalytic therapy

Less Counselor Activity

FIGURE 4.1
Counselor activity level in varying therapeutic orientations

regardless of the counselor's theoretical orientation, has been noted by many theorists and clinicians (Carkhuff, 1984; Kurpius, 1986; Rogers, 1951, 1957, 1961). Counselor empathy has been recognized as the most central dynamic in determining the quality of the therapeutic relationship (Gilbert, 1992; Kahn, 1991; Kurpius, 1986; Shea, 1988). During your internship, your ability to be empathic, to ''be there'' and understand your client's internal experiences, will be healing for your client. As a novice counselor, you will be learning to use the professional relationship between yourself and your client to help your client move forward. This therapeutic relationship may be characterized or expressed through your selection of treatment approaches or modalities (Satir, 1987).

UNDERSTANDING AND HELPING

A Holistic Perspective

The first step in choosing an appropriate treatment modality or therapeutic style involves working to understand your client from a biopsychosocial perspective. We encourage counselor interns to take note of the significance of genetic, biological, and intrapsychic forces in relation to the external influences and pressures of family, society, and world (Dacey & Travers, 1991). There is always a dynamic interplay between the client and the environment.

Consider your client's problems, behaviors, and concerns *in context* at all times, and keep in mind that an abnormal response to an abnormal situation may actually be quite normal. Your client's problematic symptom can sometimes be reframed as his or her "solution" to coping with overwhelming internal and/or external stressors. Together, you and your client can begin to understand the etiology and the purpose of the symptom in order to replace it with a more acceptable coping skill.

A Developmental Perspective

During your internship, you will also find it useful to assess your client's personality traits, concerns, and behaviors within a framework provided by personality theorists. Clients often require help in negotiating critical turning points in the lifespan or in dealing with difficult developmental tasks (Corey, 1991b). Knowledge of developmental stages allows you to understand what types of problems are likely to arise at each specific phase of your client's life cycle, so that you can be most helpful.

In addition, the developmental stages outlined by personality theorists provide a template for determining the relative health or pathology of your client (Gabbard, 1990). Behavior that is appropriate and acceptable at one stage of life may be troublesome at another stage. Gabbard (1990) provided an excellent example in the following vignette:

> Let us imagine . . . a 15-year-old boy who stands in front of the mirror blow-drying his hair for 45 minutes in order to get every hair perfectly in place. Most of us would chuckle to ourselves at this image and realize that vanity of this sort is entirely normal for a pubescent youngster. Now let us shift to an image of a 30-year-old man who spends the same amount of time in front of the mirror with a blow dryer. This visual picture is a bit more disconcerting, because such excessive self-absorption is far from the norm for a man of his age. If we now imagine a 45-year-old man engaged in the same activity, we again become a bit more charitable in our attitude because, as was the case with the adolescent boy, we understand such behavior as part of a developmental phase in the life cycle that we often refer to as midlife crisis. (pp. 369–370)*

A Multicultural Perspective

Ethnicity implies a set of prescribed values, norms, traits, communication styles, and behavioral patterns that constitute a unique cultural reality and

*From *Psychodynamic Psychiatry in Clinical Practice*, by G. O. Gabbard, p. 369. Copyright © 1990 by American Psychiatric Press, Inc. Reprinted by permission.

provide a sense of identity for members of the ethnic group (Lee, 1991). This ethnic identity colors the way the individual perceives the self and the world and provides meaning and structure to everyday activities, interpersonal relationships, and social behaviors (Gibbs & Huang, 1989). Ethnicity also determines the way people experience, express, and cope with distress, as well as how, where, and from whom they seek help (Root, 1985). Vontress (1986) pointed out that both clients and counselors are products of their cultures. Cultural factors therefore influence every aspect of the counseling process. Familiarity with multicultural issues will enable you to establish rapport and build the therapeutic relationship that is crucial for successful counseling.

Cultural sensitivity in counseling requires the knowledge base to recognize and balance universal norms, ethnic norms, and individual norms in assessing client concerns and selecting treatment modalities (Lopez, Grover, Holland, Johnson, Kain, Kanel, Mellins, & Rhyne, 1989). Cultural sensitivity helps you determine whether client behaviors, symptoms, or attitudes are actually ethnically based, rather than defensive, inappropriate, or pathological. Being aware of culturally based concepts of sickness and health, as well as behavioral norms, allows you to work with your client to set goals that are both personally and culturally acceptable. Finally, attention to cultural factors will help you select the treatment modality most congruent with your client's ethnicity, as well as enabling you to modify traditional treatment approaches as necessary to meet your client's cultural needs.

Sue and Sue (1990) wrote that certain culture-bound values present sources of conflict and misinterpretation within the context of counseling. Attention to these cultural values will help promote effective communication between counselor and client. We suggest that the following factors, based on the ideas of Sue and Sue (1990), deserve careful consideration as you work to develop a multicultural perspective:

1. Individualism versus group orientation
2. Verbal and emotional expressiveness versus reservedness
3. Valuing insight, self-exploration, or self-knowledge versus valuing "not thinking too much" or "avoiding disturbing thoughts"
4. Openness to intimacy versus reluctance for self-disclosure
5. Analytic, linear thinking versus intuitive thinking
6. Separation of mind and body versus close connection between mind and body
7. Expectation for social structure versus tolerance for ambiguity in social situations
8. Open patterns of conversation versus communication patterns organized according to social dominance and deference*

*Adapted from *Counseling the Culturally Different: Theory and Practice*, 2nd Edition, by D. W. Sue and D. Sue. Copyright © 1990, 1981 by John Wiley & Sons, Inc. Reprinted by permission.

TREATMENT MODALITIES

Psychoanalytic Therapies

Psychoanalytically oriented therapy requires a higher level of training for you and a larger investment of time and money for your client than is realistically possible during your internship. However, many important concepts and views introduced by psychoanalytic theorists form the underpinnings for other models of therapy, as well as for our current understanding of human behavior. We feel that the psychoanalytic ideas and approaches described in this section are applicable to your work and may prove to be very useful to you as a neophyte counselor.

Corey (1991a) wrote that psychoanalytic thinking provides ''a focus on the past for clues to present problems'' (p. 47). As a counselor intern, you can use this way of thinking by examining your client's developmental history to gain a more thorough understanding of your client's current level of functioning. Taking the psychosocial history is an important part of your client assessment and is discussed in more detail in chapter 5, The Clinical Interview.

The psychoanalytic concepts of transference and countertransference are also useful to you during your internship. Transference means the ''transfer of problems and feelings from past relationships onto present relationships'' (McArthur, 1988, p. 211). In counseling, the term *transference* refers to the client's unconscious tendency to experience feelings, attitudes, longings, and fears toward the counselor that were originally felt for other important people in the client's life (Kahn, 1991). For example, the client may feel strong anger toward the therapist for some seemingly inconsequential matter. Transference also includes the client's unconscious tendency to attribute attitudes to the counselor that actually were held by other important people in the client's life. For example, the client may perceive the counselor as disapproving, when in reality the counselor has no such feeling. Countertransference refers to the feelings evoked in the counselor by the client, which in fact may not be due to the client's real behavior. Countertransference has come to be more broadly interpreted to include all the reactions or feelings the counselor experiences toward the client (Kahn, 1991). Countertransference is useful to you as a counselor intern, because it may serve to help you understand the impact your client has on other people, which in turn affects the client. For example, if you find yourself feeling constantly annoyed by the client, you may understand a bit more why the client complains that people are always avoiding him or her. Self-knowledge and awareness of your own issues are absolutely essential, so that you can monitor your countertransferential feelings and be certain that they do not interfere with your effectiveness as a counselor.

Although psychoanalytically oriented therapists use transference extensively as a therapeutic means to allow clients to ''work through'' their problems, we suggest that simply understanding the concept of transference may allow you to be more helpful to your client during your internship. For example, if your client expresses hurt because you are acting ''cold'' during a session, when you in fact have warm feelings for the client, you can first examine your behavior to be certain you have not really done or said anything to give this client the feeling that you are acting "cold." Next, you can empathize and express to the client how uncomfortable it must be to have the thought that the counselor is cold. Finally, you may explore the client's feelings by asking about your counseling relationship in the present and also by inquiring whether the client has ever felt that other people have treated him or her coldly (Kahn, 1991).

The psychoanalytic term *object relations* refers to "interpersonal relations as they are represented intrapsychically" (Corey, 1991b, p. 112). Freud used the word *object* to refer to the significant person or thing that is the object of our feelings, wishes, or needs. Corey (1991b) explained that the object relations of early life are replayed throughout the life cycle, even in adult interpersonal relationships, because people unconsciously seek reconnection with their parents and therefore repeat early childhood patterns of interaction. In other words, we all tend to repeat, to some extent, the interpersonal dynamics of our family of origin. You can use this psychoanalytically oriented insight to encourage your client to become aware of repetitive patterns in relationships. Often, a client can be helped to change a potentially destructive behavior by understanding where the behavior is originating. For example, a woman who chooses emotionally distant, physically abusive men as partners, over and over again, can be helped to understand that she is recreating her own early, painful relationship with her abusive father. Together, you and this client can then work on issues such as self-esteem, anger, shame, loss, and improved coping skills, so that she does not need to be so stuck in these damaging relationships.

A familiarity with the defense mechanisms described in psychoanalytic literature will serve you well as you work to understand clients. We suggest that you take some time to review these and remember also that your client will use a defense mechanism *unconsciously* and *automatically* for protection when he or she is feeling overwhelmed, threatened, or otherwise unsafe (Clark, 1991; Satir, 1987). Your client's use of defense mechanisms is a signal for you to take notice. Perhaps the material is too difficult and you need to slow down or wait before addressing it. On the other hand, perhaps the client is avoiding the reality of painful issues and needs to be confronted or encouraged to explore the material.

Heinz Kohut (1977), a psychoanalytically oriented theorist, wrote that human beings require that three basic needs be met in order to develop a complete sense of self. These are (1) the need for mirroring, which means being valued and approved of, (2) the need for an idealized other, which means having another person to turn to for comfort and safety, and (3) the need for belonging, which means feeling like other people and being accepted as part of a group. Kohut (1977) explained that we have these needs throughout our lifetimes and that often our sense of well-being is determined by our ability to find other people, whom he termed self-objects, who can meet our needs. Although he wrote that these needs are ideally met by parents during early childhood, he suggested that our needs are ongoing and that an incomplete sense of self can be repaired even in adulthood if the needs are met by a caring other, such as a therapist. Kohut's ideas, which form the basis for his "self-psychology," have significance for you as a counselor intern in three ways. First, recognizing the underlying unmet needs that are causing your clients pain helps you to be more empathic. Gilbert (1992) offered the following example of accurate counselor empathy in recognizing and addressing a client's unmet need for an idealized other:

> The counsellor can empathize with the needs of the client in his/her wish to have some strong other come to the rescue and make things better. Thus the counsellor can recognize a client's yearning for rescue, his/her fear of abandonment and aloneness and being beyond rescue, beyond help. Hence, if a client says, "Can you help me? I feel so desperate," the counsellor should not say "Well, it's up to you." . . . Rather, the counsellor should focus on a collaborative journey, recognizing the client's need for rescue. Possible responses are, "I recognize your need for help. Let's look at your problems together and see what you find helpful. Now, what's been going through your mind?" (p. 9)*

Second, you may be more easily able to understand what sort of therapeutic relationship, response, or approach would be most healing for each client in the light of his or her unmet needs. This allows you to focus your interventions more effectively. Third, you can provide hope and a sense of empowerment to your clients by helping them understand that their needs and yearnings are universal and that there is a way to feel better, no matter how dysfunctional their families of origin were. Clients can find healing self-objects by learning to establish healthy relationships.

Cognitive-Behavioral Therapies

Cognitive-behavioral therapies are applicable in many diverse counseling settings, and the underlying concepts, as well as most of the techniques generated by these concepts, are well-suited for you to use during your internship. Cognitive-behavioral teachings stress the role of thinking in influencing our sense of well-being and in controlling our mental health. Theorists such as Ellis, Beck, and Meichenbaum maintain that emotions and behaviors are directly linked to cognitive process, so that the way we think about ourselves or the world determines how we feel and how we act (Corey, 1991a, 1991b). In cognitive-behavioral therapy, the clients' dysfunctional thought patterns and their maladaptive interpretations and conclusions about themselves and their environments are examined, tested against contradictory evidence, challenged, and replaced with more adaptive beliefs, leading to therapeutic change (Beck & Weishaar, 1989). As a counselor intern, you will often find these concepts to be useful in empowering your clients, because you will be demonstrating

*From *Counselling for Depression*, by P. Gilbert. Copyright © 1992 by Sage Publications, Inc. Reprinted by permission.

to them that their own thoughts are largely responsible for their emotional pain or behavioral problems, and therefore they can learn to help themselves to feel better and make changes by controlling their own thoughts.

Albert Ellis (1989) described his A-B-C theory in which irrational beliefs are causative factors in how we feel and act. He wrote that A represents the actual event, B represents our belief about the event, and C is the consequence of our belief, rather than the consequence of the actual event. This theoretical concept may be stated more simply as: "It depends on how you look at things" or "You can picture the glass as either half empty or half full."

Gilbert (1992) illustrated the A-B-C theory for his clients by asking them to imagine their feelings upon being awakened suddenly in the middle of the night by strange noises coming from the kitchen that sound as though an intruder is moving around. His clients usually identify feelings of fear associated with this thought. Next, he asks the clients to imagine their feelings upon being awakened by the very same noises at the same hour of the night but thinking that it is only the pet dog walking around in the kitchen. Clients identify feelings of relief or calmness, but no fear. Gilbert goes on to point out to the clients that their very different thoughts (beliefs) about the very same noises (actual event) are responsible for their differing feelings (consequence). During your internship, you may be able to use a similar story to illustrate the link between thinking and feeling for your clients, also.

As a counselor intern, you can use cognitive restructuring with your clients by helping them identify their basic irrational beliefs and automatic negative thoughts. You can help them to "think things through" in a more rational, positive manner. Often, you will be helping them explore alternative attributions for other people's behaviors, as well. For example, perhaps a client reports feeling depressed because a friend has not telephoned all week, and therefore the client assumes that the friend no longer likes him or her and feels unattractive and unlovable. As a counselor intern, you may ask your client to try to imagine other explanations for the friend's delay in calling. Could this friend have had a busy week? Been out of town? Tried to call but got a busy signal or missed finding the client at home? Simply forgot to call? Cognitive-behavioral theorists advocate using questions and a collaborative stance, so that you and the client are a team, working together. You may ask your client what the realistic possibility is that the friend actually still likes him or her, even if there was no phone call all week. You can go on to question whether the friend's liking this client actually changes the client's lovability or attractiveness. After all, the client remains the same, regardless of this particular friend's thoughts. You can go on to help the client challenge his or her assumption that self-worth depends on any other person's affection. When the client begins to understand that the friend may not have called for several possible reasons, that the friend may in fact still like the client despite not calling, that the client still has value and is not any less lovable even if the friend or any other person no longer likes him or her, the client will begin to feel less depressed. His or her maladaptive thoughts led to the depressed feelings; restructuring those thoughts leads to an improved mood.

With practice, your client can learn how to think things through in this way even outside of your counseling sessions, and cognitive-behavioral therapists call this technique "self-talk."

Another effective cognitive-behavioral technique is called catastrophizing. In practical terms, that means asking your client to verbalize the worst possible scenario and then to explore consequences and coping mechanisms. This is another way of teaching clients to think things through in a rational, logical manner, rather than being controlled by their thoughts. For example, one of our clients, a 45-year-old woman, told us she felt tense and frightened during her counseling sessions. She related this feeling to trying hard not to cry, not to let her tears out. Her irrational belief was: "I'm a mature woman and I'm not supposed to cry." Closely connected to this thought was the assumption that her counselor would not approve of her supposedly less-than-mature behavior. After empathizing with this client's feelings and the hard work she had been doing to keep from crying, we assured her that we believed crying during her counseling session was acceptable and probably helpful, rather than a sign of immaturity or weakness. We asked her to imagine what would happen if she did allow herself to cry (verbalizing the worst scenario: her crying during the session). "I probably couldn't stop, once I got started," she said (exploring consequences). However, as she thought things through, she was able to say that eventually, after a long cry, she actually would be able to stop crying (exploring coping mechanisms), and she said she also thought she might feel better (thinking things through realistically).

Corey (1991a) pointed out that possible pitfalls of cognitive-behavioral therapies are that their focus of thinking may neglect exploration of emotional issues and that they may lead to too much intellectualization by both counselors and clients. He advised that this type of therapy is not suitable for clients of limited intellectual ability. In addition, he suggested that counselors may too often tend to impose their own values and views on clients. Corey warned about "the therapist who beats down clients with persuasion" (p. 143) when using cognitive-behavioral techniques. We would also like to add that clients can easily misinterpret your efforts to challenge their irrational beliefs and see you in a punitive, disapproving, or parental role. As a counselor intern, you need to take great care to be empathic, to maintain your therapeutic relationship, and to emphasize a collaborative approach when you use cognitive-behavioral treatment modalities.

Humanistic/Existential Therapies

Humanistic/existential schools emphasize a phenomenological, experiential approach to counseling. These therapies—which draw on spiritual beliefs, existential philosophies, and psychological principles—stress the quality of the therapeutic relationship itself, rather than a set of therapeutic techniques, as an avenue toward client self-awareness and change (Corey, 1991a). The counselor's ability to grasp the client's subjective internal experience of the world and to relate to the client in an authentic manner is considered to be of paramount importance (Lewis, Hayes, & Lewis, 1986).

Carl Rogers suggested that all human beings have an actualizing tendency toward growth, health, and the fulfillment of potential, and that this tendency can be nurtured by the provision of a relationship characterized by genuineness, unconditional positive regard, and empathy (Raskin & Rogers, 1989). Rogers's teachings have two very significant implications for you during your internship. First, Rogers's belief in an inherent, universal human potential for self-actualization means that, as an intern, you do not need to be too concerned about diagnostic categories or labels in order to decide how to build a relationship with clients. You will be able to approach all clients in the same way. Second, you will be assured that even though you have relatively little clinical experience or counseling practice, you will be able to help your clients by interacting with them in an authentic, accepting, and empathic manner.

Genuineness, Rogers's first characteristic of the optimal therapeutic relationship, means being in touch with your own inner experiences and feelings and presenting yourself as a real person, rather than retreating behind a professional facade. Being genuine, or congruent, does not mean that you should reveal all aspects of your personality to your client, nor does it mean that you should express every thought and feeling. Rogers (cited in Baldwin, 1987) explained:

> The self I use in therapy does not include all my personal characteristics. Many people are not aware that I am a tease and that I can be very tenacious and tough, almost obstinate. . . . I guess that all of us have many different facets, which come into play in different situations. I am just as real when I am being understanding and accepting, as when I am being tough. To me being congruent means that I am aware of and willing to represent the feelings I have at the moment. It is being real and authentic in the moment. (p. 51)

Shea (1988) further clarified the term *genuineness* by noting that being genuine, for the clinician, means being responsive, being appropriately spontaneous, and being consistent in relating to the client.

The second aspect of Rogers's therapeutic relationship is unconditional positive regard. Rogers wrote that this concept refers to the counselor's acceptance of all the client's feelings and thoughts in a caring and nonjudgmental way: the counselor must respect and prize the client (Raskin & Rogers, 1989). As an intern, you will at some times be counseling clients whose attitudes, values, feelings, and behaviors conflict with your own beliefs and ideals. Your goal and responsibility in these situations is maintaining your unconditional positive regard for the client as a unique and valuable human being.

Empathy is the third relational quality that Rogers found to be essential in counseling. Empathy may be conceptualized as the counselor's ability to be emotionally attuned to the client, or the counselor's ability to recognize, understand, and share the client's internal world. Many authors (Gilbert, 1992; Kahn, 1991; Raskin & Rogers, 1989; Shea, 1988) have pointed out that empathic healing is actually an interpersonal, interactive process. The counselor must not only *be* empathic, but also be able to *convey* that empathic caring and interest, so that the client is able to perceive it clearly. Gilbert (1992) described a model for this empathic process: ''I understand. I show you I understand, and you understand that I have understood'' (p. 11). Kahn (1991) summarized the therapeutic value of an empathic connection between counselor and client and wrote, ''When I really *get* it that my therapist is trying to see my world the way I see it, I feel encouraged. . . . If my therapist thinks it's worth the time and effort to try to understand my experience, *I* must be worth the time and effort'' (p. 43). As a counselor intern, therefore, one of your most important tasks in helping your clients will be working hard to communicate your empathy, interest, warmth, and caring and to make certain that your clients are aware of your feelings for them.

Gestalt therapy is similar to Rogers's client-centered approach in its emphasis on client self-awareness and growth facilitated by participation in an authentic counseling relationship. Gestalt therapy is experiential and stresses client exploration and integration of thoughts, emotions, and behaviors through awareness of the here-and-now (Corey, 1991a, 1991b). The client and the counselor work to compare their individual phenomenological perspectives, exploring their differing viewpoints with the goal of increasing client awareness, insight, acceptance, and responsibility (Corey, 1991a; Yontef & Simkin, 1989).

Gestalt therapy places more emphasis on specific counseling techniques, termed *experiments* or *exercises*, than do the other humanistic/existential approaches. Many of these techniques will be useful to you during your internship, because they will direct your attention to, increase your awareness of, and encourage you to recognize the significance of (1) your clients' nonverbal behaviors and (2) your clients' use of language structure in relation to their thoughts and feelings. As a counselor intern, you will be able to understand your clients more fully and grasp the depths of their feelings and experience by attending carefully to their nonverbal signals and cues. For example, noticing your clients' muscle tension, body posture, and hand and facial movements provides you with a great deal of information. In addition, you will be able to understand and empathize with your clients' perceptions of self and world by listening closely to the words they select to express themselves. For example, the client who says, ''He makes me so depressed'' feels less powerful and assumes less responsibility than the client who says, ''I feel depressed when he does that.'' One of the Gestalt techniques that we have found to be very useful to counselor interns involves helping your client to identify and focus on the feeling, for example, anger, that is beneath his or her nonverbal behavior, such as clenched fists, and then asking your client to ''Stay with the feeling.'' As the client experiences the feeling fully, he or she often explores important issues and gains self-awareness. Another Gestalt technique you may find useful during your internship is the ''empty chair.'' Here, clients imagine someone, such as parent or spouse, to be sitting in a nearby chair. Clients then say all the things they wish they would really be able to tell that person. This directed verbalization often is a cathartic experience for the client and leads to new understandings for both client and counselor.

Corey (1991a, 1991b) cautioned that the humanistic/existential therapies focus on the present moment to such a great extent that the influence of the client's developmental history and the power of the past may be too often discounted. In addition, he suggested that the emphasis on emotional facets results in less attention to the importance of cognitive or intellectual factors in understanding clients.

ADJUNCT TREATMENT MODALITIES

Group Therapy

You will most likely be asked to facilitate or cofacilitate at least one group as part of your internship training. We suggest that you prepare for this experience by reading *The Theory and Practice of Group Psychotherapy* (1985) and, if you are working in an inpatient setting, *Inpatient Group Psychotherapy* (1983), both written by Irvin Yalom. Yalom (1985) delineated 12 therapeutic factors that are generated by the powerful interpersonal forces of groups, and he listed these in the following rank order, according to the value placed on each by individual group members:

1. *Altruism:* helping others
2. *Group cohesiveness:* belonging, being accepted
3. *Universality:* learning that I'm not the only one with these problems or feelings
4. *Interpersonal learning (input):* learning how I am perceived by others
5. *Interpersonal learning (output):* learning how to relate to others more honestly and effectively
6. *Guidance:* receiving advice from other group members
7. *Catharsis:* feeling relief by being able to express my feelings rather than keeping them hidden
8. *Identification:* finding a good role model in the therapist or other group members
9. *Family reenactment:* gaining understanding of my family of origin and insight into my old role in relating to parents, siblings, and other family members
10. *Self-understanding:* learning about and owning new parts of myself; gaining awareness of the sources of or causes of my feelings, thoughts, and actions
11. *Instillation of hope:* realizing that other people with problems like mine were able to be helped to get better
12. *Existential factors:* learning that I alone must take responsibility for my life; learning to be less concerned with small problems as I see that all people face the same larger struggles of life and death*

*From *The Theory and Practice of Group Psychotherapy*, 3rd Ed., by Irvin D. Yalom. Copyright © 1970, 1975, and 1985 by Basic Books, Inc. Reprinted by permission of Basic Books, a division of HarperCollins Publishers, Inc.

When you facilitate groups during your internship, you can focus your interventions toward (1) encouraging these therapeutic factors to come into play and (2) helping individual members to use the group's healing forces. In concrete terms, that means

- ensuring a safe place in the group by protecting members physically and psychologically
- providing appropriate structure and/or rules so that a therapeutic framework is maintained
- encouraging direct communication and interaction among group members
- modeling active listening skills
- providing positive feedback when group members take risks, work hard, show increasing self-awareness, offer support for others, etc.
- being aware of the developmental stages of group process (*Groups: Process and Practice*, by Corey & Corey, 1992, provides an excellent review.)
- asking questions such as "How many others have felt that way?" or "How many others have had a similar experience?"
- helping members become aware of and express their current feelings concerning group interaction

Bibliotherapy

Bibliotherapy involves using books or other reading material to help your clients. A book may be of the popular "self-help" variety, a booklet prepared especially for clients with a particular problem, or even a professional resource that presents relevant information without using too much jargon. You may suggest that your client read outside of your sessions, and then you may choose to discuss the books with your client during the session, or else you may read an appropriate passage together. Some groups are structured so that members take turns reading aloud, followed by a discussion of personal reactions to each section. In each situation, we suggest that you first read the book(s) yourself to be sure that the material is accurate and appropriate for your client's unique concerns and current stage in the counseling process. In addition, you can use bibliotherapy most effectively by helping your clients explore their own thoughts and feelings about the reading material. We have noticed that quite often bibliotherapy not only helps clients by providing factual information, but also offers some of the same therapeutic factors we mentioned in the section on group therapy, including universality, guidance, identification, self-understanding, and instillation of hope.

Art Therapy

Neither you nor your client needs to have any special artistic talents to use art during your counseling sessions. Art therapy is an alternative, nonverbal means of communication, which you can use in a helpful way with clients during your internship. We tell clients that "Art helps us say things that sometimes just can't be put into words." In using art with clients, keep in mind that any artwork produced during your counseling session is likely to be a very important, intensely personal representation of the client's sense of self or inner world, because feelings and issues are being stirred up by the treatment situation. Make sure to demonstrate your interest, caring, and warm acceptance of the artwork. Also, take time to explore the client's feelings and thoughts by talking about the artwork.

We suggest using simple colored markers and a 9 × 12 inch pad of white paper during your internship. These are easy to carry with you, do not require any special space or preparation, and are nonthreatening and easy for most clients to use. You may find it helpful to explain as you offer the art materials that "This is about feelings, not really about the art being good or bad." Many clients are hesitant to begin and may need some encouragement or direction. If this is the case, you can offer one or two specific suggestions about what to draw. Possible topics include: an event from the past, a dream, the client's family of origin, how the client is feeling right now, a wish, a favorite activity, a fear, an important person in the client's life, a self-portrait.

When a client draws an upsetting or frightening picture, you can help him or her feel a sense of power and control by actually changing the drawing to make it safer or less scary. For example, one client became anxious and afraid after drawing a scene depicting her childhood sexual abuse. She had drawn herself in bed, as a young girl, as her uncle began to touch her. She was able to feel safer when she offered the child in her drawing protection by adding a strong policeman to the picture and drawing handcuffs on the uncle. Another client, who was afraid to express any anger toward his father, found it safer to draw his father and then to rip the picture into small pieces, several times.

Art is a powerful therapeutic tool, and we have found it to be helpful in the following ways:

1. Art provides a cathartic experience; strong emotions are released as art is produced.
2. Art increases the connection between counselor and client, as the artwork is shared and discussed.
3. Art helps increase client self-esteem. The counselor values and accepts the client's artwork and, in so doing, demonstrates that the client is also worthy of being valued and accepted.
4. Art is empowering for the client, who is engaged in direct manipulation of his or her environment while having control of the creative process.

CONCLUDING REMARKS

During your internship, we hope that you'll value *yourself* as your most powerful means of helping your client. Recent studies (Rubin & Niemeier, 1993) have demonstrated that successful outcome in psychotherapy is positively correlated with two factors: (1) the strength of the relationship between counselor and client, and (2) the client's perception of the counselor's warmth and empathy. As a beginning counselor, your particular theoretical orientation and the specific treatment modalities you choose are not as important as your caring, your sincere efforts to understand your client, and your ability to convey to your client your empathy, respect, and desire to be helpful. Michael Kahn (1991) wrote that, as therapists, "We are not *doing* therapy the way a surgeon does surgery; we *are* the therapy" (p. 44). In this chapter, we have included an overview of some of the concepts and philosophies underlying the three major theoretical schools of counseling and psychotherapy, as well as a discussion of those aspects of the various treatment modalities we feel are most relevant to you as a counselor intern. We have limited our choices not only with regard to the constraints of the time and space of one chapter of a single book, but also with respect to your level of training as an intern. As you gain clinical experience, become more comfortable interacting with clients, encounter more difficult problems, refine your personal counseling style, and function with more autonomy in your counseling career, we encourage you to enrich your repertoire of treatment modalities and counseling methods to help your clients reach their highest levels of functioning.

References

BALDWIN, M. (1987). Interview with Carl Rogers on the use of self in therapy. In M. Baldwin & V. Satir (Eds.), *The use of self in therapy* (pp. 45–52). New York: Haworth.

BECK, A., & WEISHAAR, M. (1989). Cognitive therapy. In R. Corsini & D. Wedding (Eds.), *Current psychotherapies* (4th ed., pp. 285–322). Itasca, IL: Peacock.

CARKHUFF, R. (1984). *Helping and human relations* (Vol. 2). New York: Holt, Rinehart, and Winston.

CLARK, A. (1991). The identification and modification of defense mechanisms in counseling. *Journal of Counseling and Development, 69,* 231–235.

COREY, G. (1991a). *Manual for theory and practice of counseling and psychotherapy.* Pacific Grove, CA: Brooks/Cole.

COREY, G. (1991b). *Theory and practice of counseling and psychotherapy* (4th ed.). Pacific Grove, CA: Brooks/Cole.

COREY, M., & COREY, G. (1992) *Groups: Process and practice* (4th ed.). Pacific Grove, CA: Brooks/Cole.

DACEY, J., & TRAVERS, J. (1991). *Human development across the lifespan.* Dubuque, IA: Brown.

ELLIS, A. (1989). Rational emotive therapy. In R. Corsini & D. Wedding (Eds.), *Current psychotherapies* (4th ed., pp. 197–240). Itasca, IL: Peacock.

GABBARD, G. (1990). *Psychodynamic psychiatry in clinical practice.* Washington, DC: American Psychiatric Press.

GIBBS, J., & HUANG, L. (1989). A conceptual framework for assessing and treating minority youth. In J. Gibbs & L. Huang (Eds.), *Children of color* (pp. 1–29). San Francisco: Jossey-Bass.

GILBERT, P. (1992). *Counselling for depression.* London: Sage.

GUTHEIL, T., & GABBARD, G. (1993). The concept of boundaries in clinical practice: Theoretical and risk-management dimensions. *American Journal of Psychiatry, 150(2)*, 188–196.

KAHN, M. (1991). *Between therapist and client*. New York: Freeman.

KOHUT, H. (1977). *The restoration of the self*. New York: International Universities Press.

KURPIUS, D. (1986). The helping relationship in counseling and consultation. In M. Lewis, R. Hayes, & J. Lewis (Eds.), *The counseling profession* (pp. 96–129). Itasca, IL: Peacock.

LEE, C. (1991). Cultural dynamics: Their impact in multicultural counseling. In C. Lee & B. Richardson (Eds.), *Multicultural issues in counseling* (pp. 11–22). Alexandria, VA: American Counseling Association.

LEWIS, M., HAYES, R., & LEWIS, J. (1986). *The counseling profession*. Itasca, IL: Peacock.

LOPEZ, S., GROVER, K., HOLLAND, D., JOHNSON, M., KAIN, C., KANEL, C., MELLINS, A., & RHYNE, M. (1989). Development of cultural sensitivity in psychotherapists. *Professional Psychology: Research and Practice, 20(6)*, 369–376.

McARTHUR. D. (1988). *Birth of a self in adulthood*. Northvale, NJ: Jason Aronson.

RASKIN, N., & ROGERS, C. (1989). Person-centered therapy. In R. Corsini & D. Wedding (Eds.), *Current psychotherapies* (4th ed., pp. 155–196). Itasca, IL: Peacock.

ROGERS, C. (1951). *Client-centered therapy*. Boston: Houghton Mifflin.

ROGERS, C. (1957). The necessary and sufficient conditions of therapeutic personality change. *Journal of Counseling Psychology, 21(2)*, 95–103.

ROGERS, C. (1961). *On becoming a person*. Boston: Houghton Mifflin.

ROOT, M. (1985). Guidelines for facilitating therapy with Asian American clients. *Psychotherapy, 22*, 349–356.

RUBIN, S., & NIEMEIER, D. (1993). Non-verbal affective communication as a factor on psychotherapy. *Psychotherapy, 29(4)*, 596–602.

SATIR, V. (1987). The therapist story. In M. Baldwin & V. Satir (Eds.), *The use of self in therapy* (pp. 17–26). New York: Haworth.

SHEA, S. (1988). *Psychiatric interviewing: The art of understanding*. Philadelphia: Saunders.

SUE, D., & SUE, D. (1990). *Counseling the culturally different*. New York: Wiley.

VELLEMAN, R. (1992). *Counselling for alcohol problems*. London: Sage.

VONTRESS, C. (1986). Social and cultural foundations. In M. Lewis, R. Hayes, & J. Lewis (Eds.), *The counseling profession* (pp. 215–250). Itasca, IL: Peacock.

WALLACE, S., & LEWIS, M. (1990). *Becoming a professional counselor*. London: Sage.

YALOM, I. (1983). *Inpatient group psychotherapy*. New York: Basic Books.

YALOM, I. (1985). *Theory and practice of group psychotherapy*. New York: Basic Books.

YONTEF, G., & SIMKIN, J. (1989). Gestalt therapy. In R. Corsini & D. Wedding (Eds.), *Current psychotherapies* (pp. 328–362). Itasca, IL: Peacock.

Bibliography

BASCH, M. (1980). *Doing psychotherapy*. New York: Basic Books.

BASCH, M. (1988). *Understanding psychotherapy*. New York: Basic Books.

BECK, A., & EMERY, G. (1985). *Anxiety disorders and phobias: A cognitive perspective*. New York: Basic Books.

CASEMENT, P. (1991). *Learning from the patient*. New York: Guilford.

DRYDEN, W., & HILL, L. (1993). *Innovations in rational emotive therapy*. London: Sage.

DYER, W., & VRIEND, J. (1988). *Counseling techniques that work*. Alexandria, VA: American Counseling Association.

JOHNSON, M., FORTMAN, J., & BREMS, C. (1993). *Between two people: Exercises toward intimacy*. Alexandria, VA: American Counseling Association.

KOHUT, H. (1984). *How does analysis cure?* Chicago: University of Chicago Press.

KOTTLER, J. (1990). *On being a therapist*. San Francisco: Jossey-Bass.

NICKERSON, E. (1983). Art as a play therapeutic medium. In C. Schaefer & K. O'Connor (Eds.), *Handbook of play therapy* (pp. 234–250). New York: Wiley.

OSTER, G., & GOULD, P. (1987). *Using drawings in assessment and therapy*. New York: Brunner/Mazel.

SEINFELD, J. (1993). *Interpreting and holding: The paternal and maternal functions of the psychotherapist*. Northvale, NJ: Jason Aronson.

SELIGMAN, L. (1990). *Selecting effective treatments*. Alexandria, VA: American Counseling Association.

STORR, A. (1990). *The art of psychotherapy* (2nd ed.). New York: Routledge.

URSANO, R., SONNENBERG, S., & LAZAR, S. (1991). *A concise guide to psychodynamic psychotherapy*. Washington, DC: American Psychiatric Press.

VRIEND, J. (1985). *Counseling powers and passions*. Alexandria, VA: American Counseling Association.

WACHTEL, P. (1993). *Therapeutic communication: Principles and effective practice*. New York: Guilford.

WRIGHT, J., THASE, M., BECK, A., & LUDGATE, T. (Eds.). (1993). *Cognitive therapy with inpatients*. New York: Guilford.

CHAPTER FIVE

THE CLINICAL INTERVIEW

Many interns begin their experience by performing a "clinical interview" as part of the intake process. Note our intentional use of the word *process*. We view the intake as an important component of the therapy itself, rather than as a distinct piece of "busy work" apart from the dynamics of counseling. It has been the authors' experience that in many agencies and institutions clinicians view intakes as tedious, somewhat routine, and requiring minimal skills. As a result, interns or novice clinicians are given the task of performing the intake, which is often the first contact many clients have with the mental health system and with therapy and counseling itself. The intake, therefore, sets the tone for treatment, for sharing agency and counselor philosophy and values, and for treatment expectations.

There are many types and methods of intakes, mostly dependent on agency policy and style. Some are rather informal, calling for minimal information. Others are formalized procedures of data gathering involving the completion of several intricate forms, some by the client, others by the counselor.

We view the intake as a systematic process of information gathering and sharing usually comprising the following areas: (1) client description, (2) problem description, (3) psychosocial history, (4) mental status examination, (5) diagnostic impression, and (6) treatment recommendations (Faiver, 1988). Further, as Benjamin (1987) notes, the counselor in training should view the counseling relationship as special, thereby ensuring full concentration of energy on the client and his or her situation and unique personhood.

The client should be encouraged to ask questions and share expectations regarding treatment. Also, we suggest that you consider discussing what counseling and therapy are and are not. For example, counseling and therapy *are* a mutual process of personal exploration, growth, hard work, and anticipated problem resolution, *not* exercises in magical advice giving and unilateral direction by the counselor.

At this juncture, let's describe in detail each of the areas of the formal intake process.

CLIENT DESCRIPTION

Acquiring basic background information on your client sets the initial phase of the clinical interview. With these data, you establish a general framework by which you initiate a relationship with clients. You will soon realize this knowledge fits into your understanding of key developmental influences of individuals' lives. We suggest the following information be gathered:

Name	Religion
Address	Level of education
Birthdate	Financial data (including Medicaid, Medicare,
Gender	self-pay)
Race	Socioeconomic information
Nationality	Present job/position
Marital status	Referral source

Once the above information is collected, you may already have formulated some beliefs about the individual and the direction to take in counseling.

PROBLEM DESCRIPTION

You must encourage your clients to define their problem or concern at that particular time and be aware that explanation of their problem area is from their sole vantage point. As the interview progresses, you may discover incongruities or discrepancies that may need sorting out at some point. Delineating the scope of the problem and its effects on clients' present levels of functioning is most relevant and a key to the flow of the counseling. Inquire about their perception of the etiology of the problem and any influencing factors. Clients usually expect that you will be able to make a statement of your understanding of the problem area (Sullivan, 1970). This statement reassures clients and can only serve to enhance the development of the therapeutic alliance and subsequent therapeutic journey. Also, it can facilitate the direction of treatment. If you are not sure exactly what the defined problem is, ask and get clarification, so as not to impede the direction of the interview. Further, you can explore the history of similar problems and how they were resolved.

PSYCHOSOCIAL HISTORY

Now that the problem area has been designated, explore the client's background by taking the psychosocial history. Sullivan (1970) discusses the importance of gathering a "rough social sketch" (p. 72) of the client rather than an extensive and detailed life history. Your psychosocial history should address the developmental milestones to the present. MacKinnon and Michels (1971) believe that the psychosocial history has more diagnostic and treatment value than physical examinations or laboratory studies. Unlike Sullivan (1970), they choose to focus on an extensive developmental history from infancy to adulthood. According to these authors, being knowledgeable about the developmental milestones of each stage is relevant. In addition, they assert that the history conveys the strengths and weaknesses of the client in relation to the various interpersonal and environmental factors of his or her life.

We follow the format below. As with any outline, it should be adjusted to meet your needs.

- Developmental history—emphasizing milestones of client development
- Description of client's family—mother, father, brothers, sisters, and any significant others
- Description of early home life (happy, abusive, etc.)
- Educational history
- Occupational history
- Social activities—dating, sexual history, relationship history
- Marital history—including separations, divorces
- Health history—physical, mental, accidents, illnesses, surgeries
- Substance use, abuse, dependence
- Legal history—including probation and parole
- Present living arrangements
- Significant life events
- Lowest points of the client's life
- Abuse issues—physical, emotional, sexual
- Psychiatric history—inpatient and outpatient
- Individuals who had great influence in the client's life
- Significant environmental influences, such as school, political, religious
- Anything else we should know, but failed to ask?

We suggest that you explore in depth any specific area of history that has the most relevance to present problem areas and functioning.

MENTAL STATUS EXAMINATION

MacKinnon and Michels (1971) state that the mental status examination is "the systematic organization and evaluation of information about the client's current psychological functioning" (p. 43). The mental status evaluation is your stethoscope for understanding the client's behavioral, cognitive, and affective domains. This evaluation is a key factor in determining a diagnostic impression.

You may want to intersperse the various components of the mental status evaluation throughout your interview. This approach tends to be more comfortable for clients, because it avoids the drudgery of a long list of questions.

Faiver (1992) focuses his mental status evaluation on assessment of (1) behavior, (2) affect, and (3) cognition. We advise the intern to place equal emphasis on all three areas, but to feel comfortable, when necessary, to delve into any of them for further clarification. In addition, order of assessment should be based on personal preference and style. We shall focus briefly on each area.

Behavior

Evaluate the client's behavior as to what effect it has on present functioning. Note overall appearance, including specific physical characteristics, unusual clothing, mannerisms, gestures, tics, posture, level of self-care, level of activity, sleeping and eating disturbances, fatigue symptoms, and the client's reaction to you. What does the client's body language convey to you? Does his or her posture denote an open or closed stance? Observe whether the client can attain and maintain eye contact. Note any changes of facial expression or any other noteworthy behaviors.

Affect

Next, concern yourself with the client's feelings. Evaluate the client's overall pervasive mood. Is the client's overall feeling level dysphoric (depressed), irritable, expansive, elevated, or euphoric? Note the degree of affect. Are the client's feeling levels mild, moderate, or severe?

Cognition

Perhaps this section of the mental status evaluation is the most comprehensive because it focuses on the greatest number of variables. Note, however, that all areas of assessment in this section deal with the client's former or current thought processes. Here the counselor observes the client's alertness and responsiveness. Orientation in the three spheres (person, place, and time) is assessed. In other words, does the client know who he or she is? where he or she is? and what the date and time are? Is he or she thinking abstract or concrete? For example, is the adult client's thinking regressed and compromised? Assess any delusions (false beliefs that may or may not appear to be realistic) or hallucinations (false perceptions that may be auditory, visual, tactile, or olfactory). Are both recent and remote memory intact? Get an impression of the client's level of intelligence. The client's reality testing can be understood via his or her belief system, the level of insight into his or her problem areas, and whether the client is able to exhibit good judgment. Lastly, question the client on any ideation (thoughts) or plans (intended behaviors) to harm self or others, and note any and all responses verbatim.

As you complete the mental status evaluation, reviewing the three areas of assessment will help solidify your diagnostic impression and treatment recommendations.

DIAGNOSTIC IMPRESSION

When you arrive at this point in the clinical interview, you are cognizant of the presenting problem, possess a somewhat detailed psychosocial history, and have a good idea of the client's current mental status. The information gathered in the intake interview provides supportive documentation for the formulation of a diagnostic impression.

Counselors have mixed feelings about diagnosing. After all, is it not labeling? And labeling theory (Becker, 1963) indicates that persons often live up—or down—to a label. Also, goodness only knows in whose computer banks a diagnosis ends up. Yet, with counseling being included in the medical model, we very often must diagnose before either we or our agency gets paid. All insurers and the government require a diagnosis for reimbursement for services. Consider the great ethical and moral responsibility we have!

Licensure and certification laws in most states permit only independent practitioners (psychiatrists, psychologists, clinical counselors, clinical social workers) to diagnose mental and emotional disorders without supervision. Your responsibility is to develop a diagnostic impression based on your extensive interview. Do not minimize the importance of your diagnostic impression. Making a tentative diagnosis—an impression, not carved in stone—improves with practice, further training, and experience.

In the mental health field, the diagnostic reference is the *Diagnostic and Statistical Manual of Mental Disorders*, fourth edition, published by the American Psychiatric Association. We suggest that you view this nosology (classification system) as a cookbook, if you will. Each recipe has certain ingredients, which, in certain combinations, produce a certain product—a diagnosis. Each product has a number attached to it for ease of reporting. Diagnoses are clustered by groupings that have some similarities in symptoms. Severity is determined by several factors, including intensity, duration, and type of symptoms.

Be thorough, responsible, and accurate in your diagnosis. Choose labels carefully, because, as previously stated, labels are not always used to the client's advantage. Always consult with your site supervisor and other colleagues to verify your impressions.

You may find that your impressions change as you learn more about your client in subsequent sessions and as additional problems are discussed. It is your responsibility to note any modifications to the diagnosis in the client record.

The task of establishing a diagnostic impression must be taken seriously. No two individuals possess the same interpersonal, environmental, or genetic factors. For these reasons your ultimate decision should be individualized even though similarities may be present. We owe it to our clients to assess them in the present moment, with no preconceived notions that may serve to fit a person to a diagnosis. This affords greater objectivity and fairness.

At this point, let's take time to briefly describe the *Diagnostic and Statistical Manual* (DSM-IV) in more detail. In 1988, the American Psychiatric Association Board of Trustees appointed a task force to develop a plan to revise the then current diagnostic manual (DSM-III-R). As with previous editions, the fourth edition of the manual is compatible with the *International Classification of Diseases* manual (ICD-10), published by the World Health Organization, which has as its goal the maintenance of a system of coding and terminology for all medical disorders, including those of a psychiatric nature (American Psychiatric Association, in press). The DSM-IV is the fourth in a series of nosologies published by the American Psychiatric Association (and is actually the fifth revision). Its biopsychosocial approach attempts to consider holistically and atheoretically all possible clinical syndromes and developmental and medical variables of patients and strives to facilitate our diagnostic and treatment planning (American Psychiatric Association, in press). This is a tall order. And, certainly, the DSM-IV operates from a medical model, which can be limiting to those of us in the mental health field, considering that there are other equally compelling models, such as learning and developmental approaches to mental dysfunctions.

There are five places or categories, called axes, that are possible to address in the DSM-IV. They include the following:

Axis I	Clinical Syndromes
	Other Conditions That May Be a Focus of Clinical Attention
Axis II	Personality Disorders
	Mental Retardation
Axis III	General Medical Conditions
Axis IV	Psychosocial and Environmental Problems
Axis V	Global Assessment of Functioning

These are fully described in the DSM-IV as well as other texts. It is not our intent in this book to present a course on the DSM-IV, but merely to introduce you to the labeling system of the mental health field. We suggest that you review those books listed in the bibliography and reference sections of this chapter. The DSM-IV may very well become a constant companion and a key reference for many of you.

TREATMENT RECOMMENDATIONS

Treatment recommendations ideally are made in concert with the client. When we include the client in the decision-making process, we begin the teaching aspect of therapy: that counseling is a mutual process in which clients have valued input and equal responsibility regarding their treatment outcomes. Much depends on diagnosis, level of client acuity, client age, issues of convenience, and the expertise, skills, and theoretical orientation of the therapist. Recommendations should be based on what services would facilitate client ability to cope effectively and efficiently with current and future demands and responsibilities. Again, noting that no two individuals are exactly alike, the decision regarding treatment recommendations is as individualized as the diagnosis.

The following is a treatment recommendation checklist for assistance in choice of treatment recommendations:

1. Are client problems acute (intense) or chronic (constant over time)?
2. What limitations may impede services (finances, transportation, child care, employment, general health, time constraints, etc.)?
3. Does the seriousness of the client's situation demonstrate the need for a structured inpatient environment?
4. Can the client's concerns be handled on an outpatient basis?
5. Are there requirements from the legal system (parole board, probation department)?
6. Are we providing the least restrictive treatment environment that allows the client to address his or her problems?
7. Would the client do better in individual or group counseling?
8. Does the specific diagnosis indicate a greater success rate in a specific treatment modality?
9. Will the client likely need short- or long-term services?
10. Do you need to collaborate with other professionals (e.g., a psychiatrist, psychologist, or minister)?
11. Are there restrictions on your time and schedule that may influence your treatment?
12. Are there presenting problems outside your scope of practice necessitating transferring the client to another professional?
13. Is managed care or an insurance company directing the type and frequency of services?
14. Is your agency equipped with the trained personnel and facilities to address the client's specific problem areas (e.g., play therapy room, group therapy room)?
15. Are you able to provide any specific services the client has requested (e.g., bibliography, hypnosis)?

We attempt to assist clients to achieve their optimal level of functioning. At times, deciding on treatment recommendations may involve some creativity on our part. Even though we aspire to meet their needs, this may not always be possible. Do the best you can with what resources you have available, always operating ethically. Be candid with clients as to what you can and cannot do. For example, if you discover that neither you nor your agency can address the client's presenting problem, you have the obligation to assist that client in finding the necessary resources (ACA Code of Ethics, section B, number 12). As always, consult with your site supervisor.

CONCLUDING REMARKS

The intake is a dynamic process involving both counselor and client. Components of the process include a client and problem description, psychosocial history, mental status evaluation, diagnostic impression, and treatment recommendations. The ideal is to include the client every step of the way—as a unique and distinct partner in the therapeutic journey.

References

AMERICAN PSYCHIATRIC ASSOCIATION. (in press). *Diagnostic and statistical manual of mental disorders* (4th ed.). Washington, DC: Author.

AMERICAN PSYCHIATRIC ASSOCIATION. (1987). *Diagnostic and statistical manual of mental disorders* (3rd ed., revised). Washington, DC: Author.

BECKER, H. (1963). *Outsiders: Studies in the sociology of deviance.* New York: Free Press.

BENJAMIN, A. (1987). *The helping interview with case illustrations.* Boston: Houghton Mifflin.

FAIVER, C. (1988). An initial client contact form. In P. A. Keller & S. R. Heyman (Eds.), *Innovations in clinical practice: A source book* (Vol. 7, pp. 285–288). Sarasota, FL: Professional Resource Exchange.

FAIVER, C. (1992). Intake as process. *CACD Journal, 12,* 83-85.

MacKINNON, R. A., & MICHELS, R. (1971). *The psychiatric interview in clinical practice.* Philadelphia: Saunders.

SULLIVAN, H. S. (1970). *The psychiatric interview.* New York: Norton.

Bibliography

ADLER, A. (1964). *Problems of neurosis.* New York: Harper & Row.

ELLIS, A. (1989). Rational-emotive therapy. In R. J. Corsini & E. Wedding (Eds.), *Current psychotherapies* (4th ed., pp. 197–238). Itasca, IL: Peacock.

PIAGET, J. (1970). The definition of stages of development. In J. Tanner & B. Inhelder (Eds.), *Discussions on child development* (Vol. 4). New York: International Universities Press.

ROESKE, N. (1972). *Examination of the personality.* Philadelphia: Lea & Febiger.

CHAPTER SIX

PSYCHOLOGICAL TESTING OVERVIEW

Counselors often hesitate to include psychological testing in their assessment of clients, and many regard testing with some skepticism. Their negative attitudes may be attributed to the traditional use of testing in an exclusionary manner, to select those who will be "admitted" or "rejected," to decide who will "pass" or "fail," or to separate people into categories with labels. The historical concept and use of testing as a limiting or exclusionary tool may appear to be antithetical to the counseling profession's current orientation, which includes respecting and appreciating individual and cultural differences, viewing the client holistically, encouraging growth toward each client's optimal level of functioning, and empowering the client to take responsibility for self. Certainly, too, psychological testing has come under fire from various groups that are rightfully concerned about test bias and misuse of test results in our multicultural society (Hood & Johnson, 1991; Lewis, Hayes, & Lewis, 1986).

We would like to suggest that psychological tests can be integrated into the counseling process by using them in an inclusionary, rather than an exclusionary, way. That is, testing may be used proactively, as one important source of information within a comprehensive evaluation, to assist both counselor and client in the following ways:

- Developing a deeper understanding of the client
- Appreciating strengths and potentials
- Identifying problem areas
- Making decisions
- Setting goals
- Formulating plans
- Evaluating progress

We believe that if psychological testing is used with sensitivity and is respected as a component of the counseling process, rather than as an adjunct to counseling, it can be a therapeutic experience that helps to enhance the working alliance. In chapter 5, The Clinical Interview, we expressed a similar concept concerning the use of the initial intake assessment as an integral component of the counseling process and as an opportunity to build the therapeutic alliance.

We view psychological testing as a sort of second opinion as to what may be going on in the client's life at the moment, instead of as a final, irrefutable fact. As with any opinion, which may vary over time and may be influenced by many factors, the results of testing should always be taken with a grain of salt. (You may find it helpful to review your statistics notes for the concepts of reliability and validity at this point.) Furthermore, during your counseling internship and career, you may have access to modern computer scoring systems and methods that provide instant, lengthy, and often excellent "diagnostic" printouts based on the results of psychological testing. The caveat here is to make good use of modern technology without depersonalizing the client in the process. As counselors, we must value our professional skills, our personal intuition, and our clinical opinion as our most important therapeutic tools. View the client as a person, not a piece of interesting data!

Administering and interpreting psychological tests require advanced training, and in most settings, the responsibility for testing rests with a licensed clinical, counseling, or school psychologist. As a counselor intern, your exposure to psychological testing will vary greatly depending on your internship setting. At some agencies, interns interface only infrequently with clinicians and/or clients who are involved in psychological testing, whereas at others, interns are more actively engaged in the testing process. During your internship, your participation in this area should be closely supervised by a psychometrician or other professional who is competent in psychological testing. As a counselor intern in any setting, you should be familiar with the following basic areas of knowledge relevant to psychological testing:

- Human behavior, personality theory, and psychopathology (in order to recognize a need for testing)
- Appraisal concepts, such as power versus speed, reliability, validity, standard deviation, mean
- Types of tests available
- Specific data concerning the particular instrument selected
- Ethics and procedures related to the administration and interpretation of psychological tests
- General conditions and individual situations that may influence test results

In addition, we as counselors and counselor trainees need to keep current with technical developments in testing, so that we will be able to select the most appropriate instrument to provide information for our clients about their individual problems and concerns (Anastasi, 1992). Anastasi (1992) recommends that counselors take workshops and continuing education courses and read professional journals and other literature to stay abreast of the rapid advances in psychological testing.

We must also keep in mind that the testing experience itself affects the client and has the potential to be damaging as well as therapeutic. Communication of test results is a particularly important aspect of testing, and we suggest that when you talk with clients about their tests, you try to:

1. Help clients understand that the test is only one source of information and that the data gathered "suggests a hypothesis about the individual which will be confirmed or refuted as other facts are gathered" (Anastasi, 1988, p. 490).
2. Relate test data to the client's functioning in real-life situations, so that they are meaningful and helpful (Hood & Johnson, 1991).
3. Emphasize that the test reflects the client's functioning in the past, but that this information may be used to make changes in the future (Hood & Johnson, 1991).

Many clients are anxious about tests and may need to be reassured concerning the purpose of the instrument and the information that will be elicited. As a counselor intern, you will need to attend carefully to your clients' feelings and thoughts about each aspect of the testing process. Taking a test evokes strong feelings and thoughts in most people, which provide fertile ground for exploration and the acquisition of new insight. One counselor intern told us that she had referred her client to the agency's consulting psychologist for an MMPI to confirm a diagnostic impression of major depression and obsessive-compulsive disorder. The client was extremely upset afterward, saying, "That psychologist told me I am clinically depressed. How could he say that about me! I tried so hard on that test, too. I really thought I 'aced' it. I guess I can never do anything right." An empathic response on the counselor's part helps to relieve anxiety, promote healing and growth, and build the therapeutic alliance.

A myriad of psychological tests is extant (see Mitchell, 1985, *The Ninth Mental Measurements Yearbook*, for a complete listing with descriptions). Some will have little value to you during your counseling internship, whereas others will offer relevant information for planning the best course of action for your clients. Just as tests are designed to categorize people and their problems, behaviors, style, or preferences, psychological tests themselves can be categorized into three discrete groupings: (1) tests of intelligence, (2) tests of ability, and (3) tests of personality (Anastasi, 1988). We suggest that you take the time, either prior to your internship or early in the experience, to review these general categories of tests and their uses and purpose.

Bubenzer and his colleagues (1990) conducted a national survey of agency counselors to determine frequency of usage of psychological tests. Topping their list in terms of rate of use are

1. the Minnesota Multiphasic Personality Inventory (MMPI), now in revised form (MMPI-2) (Duckworth, 1991)
2. the Strong–Campbell Interest Inventory (SCII)
3. the Wechsler Adult Intelligence Scale–Revised (WAIS-R)

4. the Myers–Briggs Type Indicator (MBTI)
5. the Wechsler Intelligence Scale for Children–Revised (WISC-R)

Our own experience confirms that these tests appear to be the most frequently used in clinical settings. Let's take some time to examine these five tests and to understand how you, as a counselor intern, may use these tests to help your clients.

THE MINNESOTA MULTIPHASIC PERSONALITY INVENTORY (MMPI)

The MMPI is the most widely used personality inventory today and is generally considered to be "the most useful psychological test available in clinical and counseling settings for assessing the degree and nature of emotional upset" (Walsh & Betz, 1990, p. 117). It was developed in 1943 by Hathaway and McKinley at the University of Minnesota Hospitals (Anastasi, 1988) as a tool to facilitate making differential diagnoses in a population of psychiatric inpatients. Graham (1990) noted that even from the outset, the MMPI was not very successful for this purpose and cannot be used to separate people into distinct psychiatric categories. However, the MMPI has generated over 8,000 studies since it was published, and the research has been so extensive and has provided so much information that the MMPI scales are successfully used to generate a wide range of data and inferences concerning an individual's personality, emotional characteristics, and psychological problems (Graham, 1990; Hood & Johnson, 1991).

The MMPI contains 550 questions that the test-taker answers as True, False, or Cannot Say. Scores are standard T-scores with a mean of 50 and a standard deviation of 10. The cutoff point for the original MMPI is a score of 70, and generally, the higher the scores and the more elevations present, the more likely the test-taker is to have psychological problems. MMPI results are interpreted, however, according to their "profile," or overall configuration, and their "code-type," or combination of the two or three most highly elevated scale scores. An enormous amount of in-depth information about an individual's personality, symptoms, behaviors, and proclivities can be provided through analysis and interpretation of these code-types.

MMPI scores are arranged in four validity scales, which assess the subject's attitude toward the test, and ten clinical scales, which assess emotional states, social attitudes, physical conditions, and moral values (Graham, 1990). The validity scales include

1. L, or Lie, which measures responses attempting to create a favorable impression, but which are likely untrue (faking good)
2. F, or Validity, which indicates carelessness, eccentricity, malingering (faking bad), confusion, or scoring errors
3. ?, or Cannot Say, which indicates the number of questions left unanswered, and which tends to invalidate the entire test if this score is too high

4. K, or Correction, which indicates "faking good" or defensiveness if high and "faking bad" or self-denigrating attitude if low, and which is used as a factor to adjust or correct other scale scores on the MMPI

The ten original clinical scales include

1. Hs: Hypochondriasis
2. D: Depression
3. Hy: Hysteria
4. Pd: Psychopathic deviance
5. Mf: Masculinity/femininity
6. Pa: Paranoia
7. Pt: Pycasthenia
8. Sc: Schizophrenia
9. Ma: Hypomania
10. Si: Social introversion

Numerous additional scales and subscales have been developed to augment and refine information obtained from MMPI results, and the most frequently used of these are generally included in computer-scored reports. Several researchers have also developed Critical Items lists, which indicate areas warranting possible further evaluation by the mental health professional.

A revised version of the MMPI, known as the MMPI-2, was recently published. This updated instrument includes rewording to make the language more current, elimination of sexist or culturally biased terms, addition of several new scales, and restandardization using a more diverse and widely representative sample population than was used in the norming of the original MMPI (Graham, 1990). Overall, however, Graham (1990) wrote that the changes made between the two instruments are "slight" (p. 10).

During your internship, the MMPI may be useful to you in the following ways:

- You may have a source of information regarding your client's personality style or problems before your initial meeting with the client, so that you may modify your counseling interventions appropriately.
- You can use MMPI results to confirm a diagnostic impression or to "rule out" a diagnosis. This may be especially helpful in determining whether to request a psychiatric consultation for psychotropic medication, because some disorders respond better than others to psychopharmacologic intervention.
- You can discuss the Critical Items with your client as a way to explore problem areas and to help the client to disclose important thoughts and feelings. Hood and Johnson (1991) suggested that counselors always discuss Critical Items with each client, especially those related to suicidal ideation. These authors wrote that clients may assume that the counselor is already aware of all their concerns, just because a question on the MMPI alluded to an area, and may therefore not raise important issues during sessions.

- MMPI results may change over time, reflecting changes in emotional status (Graham, Smith, & Schwartz, 1986). You can therefore compare scores from tests taken at two different times to assess your client's response to life changes, stressors, or previous treatment, and you can also request an MMPI after counseling to evaluate progress.

THE STRONG–CAMPBELL INTEREST INVENTORY (SCII)

The SCII—a self-report, computer-scorable interest inventory—is the most widely used and highly respected instrument to assist clients in academic or career planning (Hood & Johnson, 1991). It assesses individual likes, dislikes, and preferences for activities and also compares the similarities and differences of individual characteristics and preferences with those of a population already employed in a particular occupation. Many sources (Anastasi, 1988; Brown & Brooks, 1991; Hood & Johnson, 1991; Walsh & Betz, 1990) stress that interest inventories, such as the SCII, evaluate interests, rather than abilities, and therefore may be used only to predict how much an individual will enjoy an occupation or an academic area of study, rather than how likely they will be to succeed or how well they will do in a given field. The SCII was first published in 1974 and represented a non-gender-biased revision that combined two earlier versions of this instrument, the Strong Vocational Interest Blank for Men and the Strong Vocational Interest Blank for Women (Walsh & Betz, 1990).

Holland's theory, characterizing personality types and the type of work environment likely to be most congruent with each personality type, underlies the structure of the SCII (Anastasi, 1988). The instrument contains 325 questions, standardized to a mean of 50 and a standard deviation of 10 (T-scores), which yield data organized into the following scale scores:

- Administrative Indexes
- Basic Interest Scales
- General Occupational Themes Scales (GOT)
- Occupational Scales
- Academic Comfort Scale (AC)
- Introversion-Extroversion Scale (IE)

The Administrative Indexes provide information that may be used to help interpret the scale scores, such as the number of questions answered and the pattern of responses. The Basic Interest Scales report the client's interest in specific areas of activity, with higher scores indicating greater interest. The General Occupational Themes Scales record the client's overall preferences related to work environment, activities, coping style, and types of people the client likes to be with and yield a three-letter Holland personality code-type, such as RIA or ASE (Walsh & Betz, 1990). These personality types, described by Holland (1985), include:

- *Realistic* (oriented toward tools or machinery)
- *Investigative* (oriented toward scientific endeavors)
- *Artistic* (oriented toward self-expressive activities)
- *Social* (oriented toward helping other people)
- *Enterprising* (oriented toward goal-directed business)
- *Conventional* (oriented toward clerical activity)

The Occupational Scales reflect similarity of the client's interests to the likes and dislikes expressed by people in various occupations. A higher score indicates that the client has more in common with people in a particular occupation. The Academic Comfort Scale rates the client's enjoyment of the academic setting, with scores below 40 tending to mean that the individual does not feel as comfortable with academic work, and scores above 50 meaning that the individual generally likes to seek out opportunities to explore theory or to do research in a field of interest. The Introversion–Extroversion Scale reflects the client's preferences for working with people (a score below 45) rather than with things (a score above 55).

Hood and Johnson (1991) stressed that clients who are self-motivated by a desire for information will benefit most from an assessment process using an interest inventory. These authors also cautioned that the SCII and other similar instruments are not suitable for use with clients who have emotional problems; these problems must be addressed prior to working on career or educational plans (Hood & Johnson, 1991). During your counseling internship, you may be able to use the SCII to help your clients in the following ways:

- You can facilitate self-exploration and self-understanding as your client examines data regarding preferences, likes, and dislikes related to occupations, academic study, and other people.
- You can assist your client with educational plans related to career goals.
- You can provide new options for occupations for your client to consider.
- You can help your client determine how much he or she has in common with other people who are employed in an occupational field.
- You can help validate your client's self-perceptions or clarify choices, and thus build self-esteem.

THE WECHSLER ADULT INTELLIGENCE SCALE–REVISED (WAIS-R)

The Wechsler Adult Intelligence Scale–Revised, published in 1981, represented several technical modifications, related to reliability and normative sample, of the Wechsler Intelligence Scale, which had been in use since 1955 (Anastasi, 1988). This instrument is used to measure adult intelligence and is organized into a Verbal Scale, consisting of six subtests, and a Performance Scale, consisting of five subtests. These subtests are alternated, one from the Verbal group and one from the Performance group, when the WAIS-R is administered (Walsh

& Betz, 1990). The scores yielded are a Verbal score, a Performance score, and a Full Scale score. The scores on the subtests are reported as standard scores with a mean of 10 and a standard deviation of 3, whereas the Verbal Scale, Performance Scale, and Full Scale scores are reported as deviation IQ scores with a mean of 100 and a standard deviation of 15 (Anastasi, 1988).

The WAIS-R Verbal Scale consists of the following subtests:

- *Information* (29 questions related to general, basic information, arranged according to level of difficulty, easiest to hardest)
- *Digit Span* (oral repetition of numbers, backward and forward)
- *Vocabulary* (35 words to be defined, arranged in order with easiest first and hardest last)
- *Arithmetic* (14 oral arithmetic problems)
- *Comprehension* (16 questions that assess understanding of meaning)
- *Similarities* (14 questions that assess ability to think in an abstract way)

The WAIS-R Performance Scale consists of the following subtests:

- *Picture Completion* (20 cards in which the client must describe what is missing from each picture)
- *Picture Arrangement* (10 cards representing parts of a story that the client must arrange sequentially)
- *Block Design* (9 cards with colored patterns that the client must reproduce using colored blocks)
- *Object Assembly* (4 puzzles that must be assembled in a limited time)
- *Digit Symbol* (a 90-second test of memory, eye-hand coordination, visual discrimination, and speed involving copying symbols for digits from a key card)

The WAIS-R may be used for differential diagnosis of psychological and mental problems, as well as for the assessment of intelligence (Anastasi, 1988; Hood & Johnson, 1991; Walsh & Betz, 1990). Variation among scores on subtests, differences among Verbal, Performance, and Full Scale scores, and overall patterns of subtest and scale scores are clinically significant. Careful analysis of WAIS-R test results by a highly trained, experienced clinician can provide information that may indicate the possibility of drug or alcohol abuse, brain damage, Alzheimer's disease, anxiety, depression, reading problems, antisocial behavior, and other psychological dysfunctions (Anastasi, 1988; Hood & Johnson, 1991). Computer scoring methods for the WAIS-R provide rapid, lengthy, and detailed descriptions of the test-taker's intelligence, learning style, areas of cognitive difficulties, personality, emotional problems, and more. We stress here, as we did in our discussion of computer-generated interpretation of the MMPI, that

1. test results comprise only one source of data about the client
2. test results must be "taken with a grain of salt"
3. the client must always be assessed in a comprehensive and holistic manner

During your internship, you may use the WAIS-R to help your client in the following ways:

- You may be able to modify or structure your counseling style and interventions to meet your client's needs more appropriately if you are aware of his or her intellectual functioning and cognitive style. For example, you may opt to use a more behavioral, less insight-oriented approach with a client who is developmentally challenged.
- You may help a client in career or educational planning.
- You may be able to help a client access beneficial community resources, such as special public school or housing programs, to which he or she is entitled, if you are aware of intellectual, neurological, or learning problems.
- You may be able to use the test scores to facilitate treatment planning by confirming a diagnostic impression or ruling out a possible diagnosis.
- You may be able to help clarify your client's problem. For example, is the source of the difficulty a cognitive, neurological, or emotional problem?

THE MYERS–BRIGGS TYPE INDICATOR (MBTI)

The Myers–Briggs Type Indicator was developed in the 1920s by Katherine Briggs and her daughter, Isabel Myers (Hood & Johnson, 1991), and it is based on the work of Carl Jung. This instrument attempts to classify individuals into one of 16 specific personality types, using a four-letter code, and provides a description of behavioral, emotional, and functional preferences. It contains 126 forced-choice questions. Hood and Johnson (1991) noted that the MBTI is popular and widely used because there are no "good" or "bad" or "higher" or "lower" scores and no dysfunctions or diagnostic categories implied by the test results. Rather, the instrument indicates tendencies toward certain characteristics and preferences but allows that individuals may also be able to use the opposite characteristic or preference.

The MBTI assesses the following personality factors:

1. *Extroversion/Introversion,* an attitude describing the flow of psychic energy outward toward the world (E) or inward toward the self (I)
2. *Sensing/Intuitive,* a perceiving function of receiving information from the outside world in a sensing (S) or intuitive (N) way
3. *Judging/Feeling,* a judging function of processing data that have been received from the outside world in a judging (J) or feeling (F) way
4. *Judging/Perceiving,* the individual's dominant means of relating to the outside world in a judging (J) or perceiving (P) way

An example of a Myers–Briggs code-type might be ENFJ, meaning that the individual is an extroverted person, with a preference for absorbing and processing information in an intuitive and feeling way, and who generally relates to the world with a judging attitude. The MBTI four-letter code attempts to present a comprehensive portrait of the way these dimensions of personality balance one another in an individual.

The instrument is relatively easy to score and interpret, and an instruction manual is included. Computer scoring methods also generate lengthy, detailed descriptions of client personality and behaviors. As always, we caution you to integrate test results as one data source within a comprehensive, evolving assessment of your client as a total human being.

The MBTI is appropriate for use in a variety of counseling settings because it provides abundant data in a nonthreatening, interesting way, and because advanced training is not necessary for the mental health professional who scores and interprets this instrument. As a counselor intern, you may use the MBTI to help your clients in the following ways:

- You may use knowledge of your own code-type as well as your client's code-type to modify your counseling interventions and improve counselor/client communication and the working alliance.
- You can help your client gain self-understanding and explore important issues by becoming aware of how his or her ways of perceiving and relating to the world affect thoughts, feelings, and behaviors.
- You can help couples and families improve interpersonal relationships by gaining an understanding of how each of their code-types and personality styles may affect communication and attitudes.
- You can help your client to choose work environments that are congruent with his or her code-type.
- You can help promote health and growth in your client by exploring the less-used functions of his or her personality style and code-type as potential coping mechanisms, new solutions to problems, alternative ways of relating to others, and new concepts of self.

THE WECHSLER INTELLIGENCE SCALE FOR CHILDREN–REVISED (WISC-R)

The Wechsler Intelligence Scale for Children was originally developed as a direct downward extension of an adult intelligence assessment instrument, the Wechsler–Bellevue Intelligence Scale (Seashore, Wesman, & Doppelt, 1950, cited in Hood & Johnson, 1991). It was further revised in 1974 (WISC-R) to include more age-appropriate questions for children and to be more multiculturally based and less gender biased. For example, the word *candy-bar* replaced *cigar* in the WISC-R (Anastasi, 1988).

The WISC-R is very similar in structure, content, and scoring organization to the WAIS-R, discussed earlier in this chapter. This instrument yields a Verbal Scale score, a Performance Scale score, and a Full Scale score, and contains 12 subtests. The subtests are administered by alternating one from the Verbal Scale and one from the Performance Scale. Two of these 12 subtests are used as substitutions or alternatives in case of a child's special needs or if the testing situation is interrupted. The subtests are expressed as standard scores with a mean of 10 and a standard deviation of 3, whereas the Verbal, Performance, and Full Scale scores are reported as deviation IQs, with a mean of 100 and a standard deviation of 15.

The Verbal Scale of the WISC-R contains the following subtests, which are very similar to those described earlier for the WAIS-R:

- Information
- Similarities
- Arithmetic
- Vocabulary
- Comprehension
- Digit span (This is an alternative subtest.)

The Performance Scale subtests are, again, very much like those we described for the WAIS-R and include the following:

- Picture completion
- Picture arrangement
- Block design
- Object assembly
- Coding (the Mazes subtest may be substituted for this.)

The WISC-R is scored, interpreted, and used in much the same way as the WAIS-R. This instrument can be used to assess children's intelligence, personality, and learning style, as well as for differential diagnosis of academic, neurological, and psychological problems. As a counselor intern, you may use the WISC-R to help your young clients in the following ways:

- You will be able to modify your counseling style and structure your interventions so that they will be more appropriate for your child client if you are aware of his or her cognitive and emotional developmental level.
- You may be able to clarify your client's difficulties as cognitive, neurological, emotional, or a combination of factors.
- You may be able to facilitate treatment planning by confirming a diagnostic impression or ruling out a possible diagnosis.
- You may be able to help parents or caretakers access beneficial school or community-based programs for which the child is eligible.
- You may be able to help parents and school personnel understand and meet the child's individual needs more effectively.
- You may be able to assist parents, caretakers, and school personnel in developing a realistic picture of the child's potentials, so that they can set appropriate goals.

CONCLUDING REMARKS

Psychological testing can serve as one valuable source of data and an important component of a comprehensive assessment of the client. The testing experience can promote self-understanding and growth and can serve to strengthen the therapeutic alliance if you, as the counselor intern, attend to your client's feelings, thoughts, and behaviors throughout the testing process. Psychological testing should be viewed as a second opinion and as a means of increasing your understanding of the client's current concerns and current level of functioning. Results should be taken with a grain of salt! We believe that test results should be used to provide more "pieces to the puzzle" in understanding your client, if you will, and should never be used as a means of disposition for any client.

References

ANASTASI, A. (1988). *Psychological testing* (6th ed.). New York: Macmillan.

ANASTASI, A. (1992). What counselors should know about the use and interpretation of psychological tests. *Journal of Counseling and Development, 70*(5), 610–615.

BROWN, D., & BROOKS, L. (Eds.). (1991). *Career choice and development.* San Francisco: Jossey-Bass.

BUBENZER, D., ZIMPFER, D., & MAHRLE, C. (1990). Standardized individual appraisal in agency and private practice: A survey. *Journal of Mental Health Counseling, 12*(1), 51–66.

DUCKWORTH, J. C. (1991). The Minnesota Multiphasic Personality Inventory-2: A review. *Journal of Counseling and Development, 69*(6), 564–567.

GRAHAM, J. (1990). *MMPI-2: Assessing personality and psychopathology.* New York: Oxford University Press.

GRAHAM, J., SMITH, R., & SCHWARTZ, G. (1986). Stability of MMPI configurations for psychiatric inpatients. *Journal of Consulting and Clinical Psychology, 54*(3), 375–380.

HOLLAND, J. (1985). *Making vocational choices: A theory of vocational personalities and work environments* (2nd ed.). Englewood Cliffs, NJ: Prentice-Hall.

HOOD, A., & JOHNSON, R. (1991). *Assessment in counseling: A guide to the use of psychological assessment procedures.* Alexandria, VA: American Counseling Association.

LEWIS, M., HAYES, R., & LEWIS, J. (1986). *The counseling profession.* Itasca, IL: Peacock.

MITCHELL, J. V. (Ed.). (1985). *The ninth mental measurements yearbook.* Lincoln, NE: Buros Institute of Mental Measurements.

WALSH, W., & BETZ, N. (1990). *Tests and assessments* (2nd ed.). Englewood Cliffs, NJ: Prentice-Hall.

Bibliography

BUROS, O. K. (Ed.). (1978). *The eighth mental measurements yearbook.* Highland Park, IL: Gryphon.

GAY, L. (1987). Selection of measuring instruments. In L. Gay, *Educational research* (pp. 125–176). New York: Macmillan.

LITWACK, L. (1986). Appraisal of the individual. In M. Lewis, R. Hayes, & J. Lewis (Eds.), *An introduction to the counseling profession* (pp. 251–277). Itasca, IL: Peacock.

OGDON, D. (1990). *Psychodiagnostics and personality assessment: A handbook.* Los Angeles: Western Psychological Services.

WALLACE, S., & LEWIS, M. (1990). *Becoming a professional counselor.* New York: Sage.

CHAPTER SEVEN

PSYCHOPHARMACOLOGY FOR COUNSELOR INTERNS

Over the past 30 years, scientific investigation into the biological and genetic underpinnings of mental illness has resulted in the development of many new medications, as well as the discovery of new uses for existing medications, which have changed the face of the mental health field (Schatzberg & Cole, 1991). These psychiatric (also called psychotropic) drugs provide hope and relief to those suffering from serious psychiatric disorders by alleviating intrapsychic pain, improving cognitive functioning and interpersonal skills, and reducing those behaviors and symptoms that result in social stigmatization (Gitlin, 1990). The use of psychiatric drugs has resulted in far fewer and briefer hospitalizations for individuals with severe mental illness over the past several decades (Yudofsky, Hales, & Ferguson, 1991). Research indicates that psychotropic medications are extremely helpful in the treatment of a wide range of emotional and mental problems, including major depression (Heninger, Charney, & Sternberg, 1983; Kane, Quitkin, Rifkin, Ramos-Lorenzi, Nayak, & Howard, 1982), bipolar disorder and mania (Fawcett, 1989), anxiety (Noyes, 1985), panic disorders (Ballenger, 1986), schizophrenia (Delva & Letemendia, 1982), obsessive-compulsive disorder (DeVeaugh-Geiss, Landau, & Katz, 1989), addictions (Extein & Gold, 1988), attention-deficit hyperactivity disorder (Biederman, 1988; Cantwell, 1985), chronic pain (France, Houpt, & Ellinwood, 1984; Raft, Davidson, Mattox, Mueller, & Wasik, 1979), impulsive or antisocial behavior (Salzman, 1988), dementias (Raskind, Risse, & Lampe, 1987), sleep disturbances (Kales, Soldatos, Bixter, & Kales, 1983), borderline personality disorder (Cowdry, 1987), and eating disorders (Goldbloom & Olmstead, 1993). Psychotropic drugs are intended to be used to help restore individuals to their usual state of functioning and feeling, rather than to gain improvements beyond their normal levels of functioning (Schatzberg & Cole, 1991; Yudofsky et al., 1991). "Although they exert profound and beneficial effects on cognition, mood, and behavior, they often do not change the underlying disease process, which is frequently highly sensitive to intrapsychic, interpersonal, and psychosocial stressors" (Schatzberg & Cole, 1991, p. 2). Outcome research studies indicate that in the treatment of emotional problems a concomitant use of psychiatric medication with psychotherapy often provides more improvement and greater benefits than treatment with either drugs or therapy alone (Luborsky, Singer, & Luborsky, 1975; Smith, Glass, & Miller, 1980).

All physicians, osteopaths, and dentists are licensed to prescribe medication, and many psychiatric drugs are actually prescribed by doctors who specialize in areas other than psychiatry, for example, family practitioners or pediatricians. Psychiatric drugs therefore comprise a large proportion of the most commonly prescribed medications today, and, in fact, one brand of anti-anxiety agent (Xanax) was recently ranked third among the top ten most frequently used medicines (Salzman, 1991). Because psychotropic medications are employed so often, particularly in the clinical mental health field, counselors need to be aware of the indications (uses), contraindications (inappropriateness), and side effects of these drugs.

As a counselor intern, you may be the mental health professional who interacts face-to-face with the client most often. Your responsibility includes being alert to those clinical situations that indicate the need for a psychiatric consultation and possible psychopharmacologic intervention. During your internship, you may also have the task of monitoring client behaviors and symptoms, understanding which of these can be influenced by medications, and subsequently communicating your observations to the psychiatrist or to your supervisor. The following symptoms or situations suggest that the client would benefit from an evaluation by a psychiatrist to determine the appropriateness of treatment with psychotropic drugs (please note that psychiatric evaluation is indicated in many other clinical situations, as well; consult your supervisor and follow your agency policies and guidelines):

- Significant signs of depression, such as changes in appetite, sleep problems, low energy level, extreme fatigue, concentration difficulties, hopelessness, psychomotor retardation, and suicidal ideation
- Hallucinations
- Delusions, paranoia, impaired thinking, or mental confusion
- Signs of a manic episode, including abnormally elevated or irritable mood, pressured speech, increased energy and psychomotor agitation, and excessive involvement in pleasurable activities that have a high potential for destructive consequences
- Obsessive thoughts and/or compulsive or ritualistic behaviors
- Moderate to severe anxiety, or panic attacks
- Family history of mental illness

You may also need to educate your clients concerning psychotropic medications that have been prescribed for them, and you may need to explore in therapy any special emotional significance of the medication to each client in order to encourage compliance or to discourage excessive dependency, when necessary. Psychopharmacologic intervention may be understood as a treatment modality that, although it is biologically or medically based, encompasses many of the same issues and dynamics as do other psychotherapeutic treatment modalities (Gabbard, 1990). For example, clients may experience a transferential relationship with the doctor who prescribes the medicine, and these feelings may influence their acceptance or rejection of the drug. Perhaps the resistant or noncompliant client fears being labeled as "sick," worries about "losing control," or believes it is a sign of "failure" if psychiatric medication is prescribed. On the other hand, perhaps the client demands too much medication as a way of asking for proof of the doctor's affection or caring, as a means of "numbing" to avoid painful issues or feelings, or as an avenue for maintaining control of the treatment situation. As a counselor intern, you should be listening to your clients' concerns about their medication just as seriously, carefully, and empathically as you possibly can. We have found it helpful to reassure clients about the following:

- Often several different medicines and doses must be tried out to find just the right one that will provide the greatest benefit with the fewest side effects.
- Many psychotropic medications require two weeks or more to reach therapeutic levels in the blood.
- Sometimes side effects and other medication-related symptoms increase initially when a drug is started, but then decrease or disappear with time. Patience helps.
- The purpose of psychiatric medication is to help people feel more like their real selves, at their best, not to change people's personalities.
- Taking a psychiatric medication does not mean you are crazy.
- Often the psychiatric medicine just helps to "take the edge off," so that you can focus on working in therapy to improve your emotional health.

Clients' worries about psychotropic drugs and their side effects should be considered important issues and addressed respectfully. *Always* communicate closely with the psychiatrist, your supervisor, the psychiatric charge nurse, or whomever your agency designates responsible, concerning your clients who are taking psychiatric drugs, specifying any difficulties, side effects, new symptoms, improvement, or lack of improvement in the clients' functioning. We also suggest documenting all this information in writing in your clients' charts and records.

Principles of psychopharmacologic intervention are based on attending to six factors (Davis, 1985):

1. Careful diagnosis of the emotional or mental disorder
2. Selection of appropriate target symptoms
3. Consideration of all possible benefits compared to possible risks (including drug and food interactions, the client's medical status, likelihood of compliance, and side effects)
4. Determination of correct dosage
5. Monitoring of response
6. Possible adjustment of dose and medication*

*From *Clinical Psychiatry,* edited by H. S. Sudak. Copyright © 1985 by Warren H. Green, Inc. Reprinted by permission.

Psychiatric medication may be classified into four groups or clas tidepressants, antipsychotics, antianxiety agents, and mood stabilizer following sections, we discuss the various psychotropic medications by class and clinical use, provide the trade and chemical names of the m monly prescribed drugs in each category, and list possible side effects information, and cautions for each class of drug.

ANTIDEPRESSANTS

Antidepressants are used to treat the symptoms of depression and should not be considered "mood elevators," because an individual who is not depressed will not experience a "high" or any other effect on mood (Yudofsky et al., 1991). Some types of antidepressants have also been used successfully in the treatment of psychogenic pain, eating disorders, panic, phobias, obsessive-compulsive disorder, anxiety, migraines, insomnia, withdrawal from chemical addictions, attention-deficit disorder, post-traumatic stress disorder, and childhood enuresis, although the FDA has not approved the use of these drugs in all these cases.

Antidepressants have little or no potential for promoting drug dependence or addiction. The main concerns in using these medications are their troublesome side effects, the possibility of serious interactions with food or other drugs when using certain classes of antidepressants, and the lethality of overdose, whether by accident or in suicide attempts.

In a relatively new process called "augmentation therapy," antidepressants are prescribed in combination with other drugs to enhance their therapeutic effects (Salzman, 1991; Yudofsky et al., 1991). Medications used to increase the activity of antidepressants include very small amounts of lithium, commercially manufactured thyroid hormone, and stimulants (Schatzberg & Cole, 1991).

There are three types of antidepressants: (1) heterocyclics (also called tricyclics), (2) monoamine oxidase inhibitors (usually called MAOIs), and (3) serotonin-re-uptake inhibitors (SRIs). We will discuss each of these classes in more detail below.

Heterocyclics

Trade name	Chemical name
Adapin	doxepin
Amitril	amitriptyline
Anafranil	clomipramine
Asendin	amoxapine
Elavil	amitriptyline
Endep	amitriptyline

Trade name	Chemical name
Janimine	imipramine
Ludiomil	maprotiline
Norpramine	desipramine
Pamelor	nortriptyline
Pertofane	desipramine
Sinequan	doxepin
Surmontil	trimipramine
Tofranil	imipramine
Vivactil	trimipramine

Possible side effects   Anticholinergic effects (including blurry vision, constipation, dizziness, dry mouth, and urinary problems), skin rash, sedation, sexual dysfunctions, increased appetite and weight gain, tachycardia (racing heartbeat), flushing, blood and liver disorders, worsening of glaucoma, increased anxiety. Many of these disappear by themselves with time or may be treated symptomatically, such as sipping beverages or sucking hard candies for dry mouth and increasing bran or bulk in diet for constipation.

Special information   Heterocyclic antidepressants are considered safe, effective medications for the treatment of depression; approximately 80% of people with major depression improve with this type of drug (Yudofsky et al., 1991).

Cautions   The risk of fatal overdose is very high with heterocyclics, and because they are used to treat a population who may be feeling suicidal, psychiatrists often prescribe only a few pills at a time. However, counselor interns need to be aware that some clients may "save up" medication to accumulate a lethal dose, and therefore careful monitoring of suicidal ideation and feelings of worthlessness and hopelessness is necessary.

Monoamine Oxidase Inhibitors (MAOIs)

Trade name	Chemical name
Eldepryl	selegiline
Eutonyl	pargyline
Marplan	isocarboxazid
Nardil	phenelzine
Parnate	tranylcypromine

Possible side effects   Dizziness, constipation, urinary problems, dry mouth, weight gain, insomnia at night and sleepiness during daytime, muscle cramps, orthostatic hypotension (light-headedness due to low blood pressure when standing or sitting up).

Special information MAOIs are used to treat depression when the heterocyclics have not been helpful; they are also especially effective in treating panic attacks, depressions with anxiety, and atypical depressions (depressions that have more unusual symptoms occurring than are typically found in major depression, such as irritability rather than sadness).

Cautions Most surgeons and anesthesiologists advise that patients stop taking MAOIs before undergoing surgery because of possible blood pressure complications (Schatzberg & Cole, 1991). Another major concern in the use of MAOIs is the severe, often fatal food and drug interactions that can be precipitated. Clients who are taking MAOIs need to be careful to control their diets and to avoid foods that are high in tyramine, as well as to avoid using several types of prescription and over-the-counter drugs. Therefore, MAOIs should not usually be prescribed for clients who have memory problems or are confused. The list of foods and drugs that must be avoided when taking MAOIs includes the following (please note that these are only partial lists of problem foods and medications):

Foods
Cheese (including cheddar, Gruyère, American, and others)
Beer, red wine, sherry, hard liquor, champagne
Beans
Lox, smoked fish, pickled herring
Raisins, figs
Sausage, liver
Avocado, bananas
Chocolate, cocoa
Yeast
Yogurt

Medications
Local anesthetics, such as Novocain
Decongestants
Ritalin
Demerol
Dexedrine
Antihistamines

Serotonin Re-Uptake Inhibitors (SRIs)

Trade name	*Chemical name*
Desyrel	trazedone
Paxil	paroxetine hydrochloride
Prozac	fluoxetine
Zoloft	sertraline hydrochloride

Possible side effects   Serotonin re-uptake inhibitors (SRIs) are the newest class of antidepressant medications. They increase the time that serotonin, the body's natural mood elevator, remains in the brain before being reabsorbed. In general, SRIs cause fewer anticholinergic side effects (dry mouth, blurred vision, constipation, urinary problems) than the heterocyclic antidepressants, and SRIs are not associated with the serious food or drug interactions that may be precipitated with the MAOIs. The most common side effects include dizziness, anxiety, insomnia, and stomach upset. Trazedone (Desyrel) may sometimes cause sustained erection in men that is serious enough to require emergency medical or surgical intervention. In addition, this drug has been reported to cause severe dizziness and fainting in some patients if taken on an empty stomach (Schatzberg & Cole, 1991).

Special information   Prozac has received a great deal of attention in the media because of reports that depressed patients who were treated with this drug became more suicidal as a result of taking it. However, Prozac is considered to be very safe and highly effective by most experts in psychopharmacology, and it is widely used in the treatment of depression and, at times, obsessive-compulsive disorder, phobias, and panic attacks (Schatzberg & Cole, 1991; Yudofsky et al., 1991). The SRIs are not as toxic as the heterocyclics when taken in large amounts, and therefore patients are not generally at as great a risk for fatal overdose, whether by accident or in a suicide attempt.

Cautions   People taking SRIs should not drink any alcohol at all and should check with their medical doctors before using other drugs. Fatal drug interactions between SRIs and the MAOIs have been reported, so that these medications should not be used at the same time. Furthermore, clients should not begin taking MAOIs sooner than 5 weeks after stopping SRIs, because medication may remain in the blood that long after dosage has been discontinued (Schatzberg & Cole, 1991).

ANTIPSYCHOTICS

Antipsychotics (also called neuroleptics) are used to treat psychosis, a condition where the client's ability to distinguish between what is real and what is not real is impaired. Psychosis includes delusions (for example, paranoia), hallucinations, thought disorders, and sometimes behavioral manifestations such as combativeness, agitation, and aggression. Psychosis may be due to schizophrenia and other psychiatric disorders (including acute mania and serious depression), brain tumors, alcohol, reactions to drugs and other toxic chemicals, head injury, stroke, infection, medical illness, and as other problems. Antipsychotic medications are also used to treat Tourette's syndrome, debilitating nausea, schizoaffective disorder, borderline personality disorder, and agitation due to organic disorders. They are very effective when combined with antidepressants in the treatment of psychotic depression. Antipsychotics often provide dramatic relief for clients and their families by removing or reducing the clients' intrusive, possibly frightening hallucinations, faulty beliefs, and disorganized thinking, and lessening anxiety, hostility, and agitation, allowing them to communicate with others and to feel less isolated.

Antipsychotic medications are generally grouped according to class of chemical structure and also according to how much of the medication must be taken to achieve a therapeutic outcome. Thus, antipsychotics may be categorized as low potency (requiring more medication), intermediate potency (requiring a medium amount of the drug), or high potency (requiring less medication).

Trade name	Chemical name	Potency
Clozaril	clozapine	intermediate
Haldol	haloperidol	high
Loxitane	loxaoine	intermediate
Mellaril	thioridazine	low
Moban	molindone	intermediate
Navane	thiothixene	high
Orap	pimozide	high
Prolixin	fluphenazine	high
Stelazine	trifluoperazine	low
Thorazine	chlorpromazine	low
Tindal	acetophenazine maleate	intermediate
Trilaphon	perphenazine	intermediate

Possible side effects   Anticholinergic effects (including dry mouth, constipation, blurred vision, and urinary difficulties) and sleepiness may resolve over time and may be treated symptomatically, except in the case of urinary retention, which requires medical care. Additional possible physiological side effects of antipsychotic drugs include itching, rash, restlessness, confusion, flushing, sweating, endocrine changes and breast enlargement in both males and females, orthostatic hypotension (blood pressure changes due to changes in posture and position), nasal congestion, jaundice, diarrhea, sexual dysfunctions, vomiting, sensitivity to light, weight gain, difficulty swallowing, and headaches (Davis, 1985; Salzman, 1991).

Antipsychotic medications may cause several other adverse side effects, as well, such as tardive dyskinesia, pseudoparkinsonianism (akinesia), acute dystonic reaction, motor restlessness (akathisia), malignant neuroleptic syndrome, and cardiac problems. These terms will be defined and discussed below.

1. *Tardive dyskinesia* involves involuntary, abnormal motor activity, especially sticking out the tongue, smacking or licking the lips and mouth, blinking, grimacing and frowning, moving the limbs, and twisting the neck and trunk. These motions may become severe enough to interfere with breathing (Yudofsky et al., 1991). Tardive dyskinesia may be permanent and irreversible once it develops, so physicians strive to detect this syndrome early and to prevent it from

occurring by using the lowest possible doses of antipsychotic medications for the shortest possible periods of time.

2. *Pseudoparkinsonianism (akinesia)* refers to a condition due to antipsychotic medications where the face has a masklike, unmoving expression, and the client manifests many of the signs of Parkinson's disease. Often, there is drooling, tremor of the hands, stiffness of the joints, a shuffling gait, and retardation of body movement in general (Gitlin, 1990). Akinesia may occur within a few days to a few weeks after a client begins taking antipsychotics, and it may be treated successfully with appropriate medication.

3. *Acute dystonic reaction* is a sudden, acute contraction of a muscle group, causing distressing tightening or twisting. Frequently, this occurs within hours of taking the medication and involves the spinal cord, the neck, or the eye muscles (Yudofsky et al., 1991). Acute dystonic reaction responds well to emergency medical treatment.

4. *Motor restlessness (akathisia)* occurs in as many as 40% of clients after only a single dose of Haldol and in 75% of clients who take this antipsychotic for a week (Gitlin, 1990). Clients with this condition feel that they can get relief from their discomfort only by jumping up and down, swinging their legs, or moving constantly.

5. *Malignant neuroleptic syndrome* is an uncommon but potentially life-threatening complication that is most likely to occur in three situations: (a) upon initiation of antipsychotic drug treatment, (b) when dosages are suddenly increased, or (c) when high doses are being used (Gitlin, 1990). This condition is a serious medical emergency requiring immediate hospitalization. Symptoms include mental confusion or coma, rigid muscles, fever, rapid pulse, blood pressure abnormalities, and breathing difficulty.

6. *Cardiac problems* related to the use of antipsychotic medications include unusually high or low blood pressure, stroke, fainting, and angina (chest pain).

Special information   The only antipsychotic medication that does not ever result in tardive dyskinesia is Clozaril (clozapine), a fact that underscores how useful this drug is for individuals with severe, chronic mental illness, such as schizophrenia. Thousands of families have been relieved of needing to choose between their loved ones suffering psychotic symptoms without medication or their suffering tardive dyskinesia with antipsychotic medication. However, Clozaril has been associated with an uncommon, but potentially fatal blood complication, and therefore weekly blood tests are required for all individuals taking this drug. Unfortunately, the cost of Clozaril plus the blood tests is currently about $9,000 per year, per patient (Schatzberg & Cole, 1991), meaning that many people who could benefit from this medication cannot afford it.

Cautions   Antipsychotic medications are often extremely sedating, so that people taking these drugs may be too drowsy to drive, to operate machinery, or even to smoke cigarettes safely (Yudofsky et al., 1991). Antacids interfere with the absorption of antipsychotic medication and therefore should not be taken sooner than two hours before or after the medication. Alcohol must be completely avoided, and patients should check with their doctors before taking all other prescription as well as over-the-counter drugs because of possible adverse interactions. Antipsychotic medications often limit the body's ability to regulate temperature, so that patients need to avoid lengthy exposure to hot weather conditions, to refrain from overexercising, and to drink plenty of fluids to keep well hydrated.

ANTIANXIETY AGENTS

Most people feel anxious sometimes, and anxiety that interferes with daily functioning may be considered "the most widespread of all psychological problems" (Yudofsky et al., 1991). Debilitating or severe anxiety is likely to be associated with psychiatric disorders such as mania, post-traumatic stress disorder, generalized anxiety disorder, panic disorders, agoraphobia, obsessive-compulsive disorder, agitated depression, and performance anxiety. Anxiety may also be due to the ingestion of caffeine and other foods and drugs, as well as to a variety of medical conditions, including hypoglycemia (low blood sugar), hypertension (high blood pressure), asthma, some types of tumors, and thyroid disorders. Therefore, a thorough physical examination is always necessary in the evaluation of anxiety.

Antianxiety agents are used to alleviate the physical and psychological symptoms of anxiety. These symptoms may include feelings of nervousness, irritability, apprehension, rapid or pounding heartbeat, sweating, chills, fatigue, gastrointestinal upset, dryness in the mouth, difficulty swallowing, dizziness, sleep disturbances, headache, and muscle aches and pains. Antianxiety medications are also used as muscle relaxants, in withdrawal from alcohol, to control seizures, as preanesthetic agents in surgery, and sometimes as sleep aids in low doses.

Thirty years ago, most antianxiety agents belonged to a class of drugs known as barbiturates, which were not only highly addictive and very sedating, but also extremely dangerous because of their potential for lethal overdose. In the early 1960s, a new class of drugs, the benzodiazepines, were introduced, and these were somewhat less likely for lethal overdose and also somewhat less addictive. However, even the benzodiazepines may be habit forming, depending on the size of dose, length of treatment, and particular patient. In addition, they often cause people to feel quite sleepy or "spaced out" and are dangerous when taken with alcohol. Another newer class of medication, buspirone (Buspar), which was originally developed as an antipsychotic, has recently been recognized as a safer, nonaddictive, nonsedating antianxiety agent. Antihistamines and beta-blockers are also used safely and effectively to relieve anxiety.

Barbiturates

Trade name	Chemical name
Amytal	amobarbital
Butisol	butabarbital
Nembutal	pentobarbital
Phenobarbital	phenobarbital

Possible side effects Ataxia (problems with balance), coordination difficulty, slurred speech, sedation and sleepiness, increased hyperactivity and impulsive behavior in children and adults who have attention-deficit disorder, serious respiratory problems (including spasms of the larynx).

Special information Phenobarbital is still sometimes used in the treatment of epilepsy to control seizures and as a surgical anesthetic, but otherwise this class of antianxiety medication has largely been replaced by the newer antianxiety agents. Amytal (amobarbital) is the drug that the media have labeled ''truth serum,'' because it frequently allows patients to speak more freely or to remember repressed material while under its influence.

Cautions Barbiturates are extremely addictive and can cause severe and possibly life-threatening symptoms during withdrawal from the drugs. Fatal overdose occurs frequently. Clients must not consume any alcohol when taking barbiturates.

Benzodiazepines

Trade name	Chemical name
Ativan	lorazepam
Centrax	prazepam
Librium	chlordiazepoxide
Paxipam	halazepam
Serax	oxazepam
Tranxene	clorazepate
Valium	diazepam
Xanax	alprazolam

Possible side effects Sedation, nausea, hypotension (low blood pressure), ataxia (balance problems), dizziness, coordination difficulty, weakness, fatigue, memory problems, rash, itching, constipation, insomnia, sexual dysfunction. Aggressive or angry behavior, serious depression, and mania may also occur in rare cases as a result of taking these medications.

Special information This class of antianxiety medication is addictive and will cause adverse psychological and physical symptoms if the drug is stopped too suddenly. Physicians usually prescribe decreasing doses and taper the drug gradually when the medication is no longer needed. Withdrawal symptoms may include rebound anxiety (return or increase of original symptoms of anxiety), panic attacks, sleep disturbance, depression, memory problems, tremor, hypersensitivity to sound and light, nausea and vomiting, sweating, rapid pulse, feelings of depersonalization, and, in rare cases, grand mal seizures or psychotic episodes.

Cautions   People taking this class of antianxiety medication should not drink any alcohol. In addition, the benzodiazepines can cause potentially dangerous interactions when combined with other types of drugs, so that patients should consult with their doctors before taking any other medications. People taking a benzodiazepine are often advised not to drive or operate machinery due to the sedative effect of the drug. This class of antianxiety medication is contraindicated for people who have had problems in the past with chemical dependency; who have multiple sclerosis, stroke, or neurological problems (Yudofsky et al., 1991); or who have a personality disorder (Schatzberg & Cole, 1991).

Buspirone

Trade name	*Chemical name*
Buspar	buspirone

Possible side effects   Nausea and other gastrointestinal upset, dizziness, nervous tension, headache, sleep problems.

Special information   Buspar provides effective relief of anxiety, except panic disorders (Schatzberg & Cole, 1991). It is not addictive, does not interact adversely with alcohol or other drugs, is not sedating, and does not impair coordination or balance. Buspar requires a waiting period of one to three weeks on the drug before patients begin to realize a therapeutic effect, however, and this is too long a time for some people who are suffering intense anxiety. In addition, there have been some reports that Buspar is not as effective in people who have previously taken one of the diazepines for anxiety (Schatzberg & Cole, 1991).

Cautions   Buspar should not be used in combination with the MAOIs, because these medications can interact to cause a dangerous rise in blood pressure.

Antihistamines

Trade name	*Chemical name*
Atarax	hydroxyzine
Benadryl	diphenhydramine
Vistaril	hydroxyzine pamoate

Possible side effects Anticholinergic effects (dry mouth, constipation, urinary problems, dizziness), sedation, and drowsiness.

Special information Antihistamines are not quite as effective as the other antianxiety agents in relieving symptoms of anxiety, but are frequently prescribed because of their highly sedative properties. In addition, they are often used because they are not addictive and have few serious side effects.

Cautions People taking antihistamines should not drink alcohol and should check with their doctors before taking other medications. The sedative effects may interfere with the ability to drive, operate machinery, or follow the usual daily routine safely.

Beta-Blockers

Trade name	Chemical name
Catapres	clonidine
Inderal	propanolol
Tenormin	atenolol

Possible side effects Fatigue, dizziness, hypotension (low blood pressure), sexual dysfunction, dry mouth, depression.

Special information Beta-blockers help relieve the physical sensations of anxiety, including sweaty palms, upset stomach, and pounding heart. They work quickly and are therefore prescribed for performance anxiety before stressful or important events. Beta-blockers are not as effective for other types of anxiety, and their use is generally limited to treating performance anxiety. However, they have many medical uses, such as in hypertension, angina, cardiac arrhythmias (irregular heartbeat), migraines (Schatzberg & Cole, 1991), tremor, and akathisia (motor restlessness) (Gitlin, 1990).

Cautions Beta-blockers and the antipsychotic drug Thorazine increase the blood level of both drugs when they are taken together, which increases the possibility of adverse side effects from both drugs, as well. This potentially increases the risk of tardive dyskinesia (a side effect of Thorazine, which we discussed in our section on antipsychotics). Beta-blockers may also be contraindicated in patients with impaired liver or kidney function.

MOOD STABILIZERS

Mood stabilizers are used in the treatment of mania and in bipolar (manic-depressive) disorder to diminish and prevent the emotional highs and lows of this illness. Mania may be caused by medications, such as steroids, antidepressants, and others, as well as by certain medical conditions, such as epilepsy, thyroid disorders, and multiple sclerosis (Yudofsky et al., 1991). Symptoms of mania or the manic phase of bipolar disorder may include rapid or pressured speech, feelings of grandiosity, decreased need for sleep, racing thoughts, increased activity level or psychomotor agitation, and excessive involvement in potentially harmful activities, such as indiscriminate shopping sprees. Mood stabilizers also have been used effectively in cases of borderline personality disorder, alcoholism, chronic schizophrenia, schizoaffective disorder, impulsive anger, affective lability, and cyclothymia (Schatzberg & Cole, 1991). Lithium salts and anticonvulsants are the two general types of drugs that are most often used to control mania and to stabilize mood in bipolar disorder.

Lithium Salts

Trade name	Chemical name
Cibolith	lithium citrate
Eskalith	lithium carbonate
Lithane	lithium carbonate
Lithobid	lithium carbonate
Lithotabs	lithium carbonate

Possible side effects   Tremor of the hands, forgetfulness, nausea, diarrhea, increased urination, kidney problems, weight gain, rash, itching, hair loss, thyroid disorders, dizziness, balance problems, sleep disturbance.

Special information   There is a narrow therapeutic margin of safety for lithium, and if blood levels rise too high, this drug is toxic and can be fatal. Frequent monitoring, through blood tests, is necessary. Furthermore, because lithium is excreted directly through the kidneys, anything that changes the body's water volume also changes blood lithium levels. Patients taking lithium are advised to drink plenty of fluids and to avoid exercising to the point where they become dehydrated.

Cautions   Lithium may cause drug interactions when used with other medications, including antibiotics, asthma medicines, antipsychotics, and diuretics. People taking lithium should consult their doctors before taking any combination of medications. Alcohol is prohibited. Lithium may be contraindicated in people who have a history of high blood pressure, glaucoma, aplastic anemia, or kidney, liver, or thyroid disorders (Yudofsky et al., 1991).

Anticonvulsants

Trade name	Chemical name
Depakene	valproic acid
Depakote	valproic acid
Klonopin	clonazepam
Tegretol	carbamazepine

Possible side effects Severe or sometimes fatal blood and liver disorders, sedation, tremor, balance problems, hair loss, weight gain, coordination difficulty, possible increased anger (Schatzberg & Cole, 1991), nausea and vomiting, blurred vision, dry mouth.

Special information Because of the potentially serious or lethal blood and liver disorders that may occur with these medications, patients need to be aware of and watch out for the early warning signs. These include easy bruising, bleeding, fever, and sores in the mouth. Frequent blood tests are necessary, because an abnormal blood count is another warning sign.

Cautions These mood stabilizers cause adverse interactions with many other types of drugs, so patients are advised to check with their doctors before taking any combination of medications. Alcohol increases the sedative effect of anticonvulsants and should therefore be used cautiously. Yudofsky et al. (1991) advised that anticonvulsants should not be used for children under age 6 or for people who have a history of liver disease, aplastic anemia, allergy to medications, narrow angle glaucoma, or bone marrow depression.

CONCLUDING REMARKS

Psychotropic medications offer relief for a wide spectrum of psychological problems, comprise an integral aspect of treatment of some emotional and mental disorders, and provide an important adjunct to psychotherapy. However, these medications must be prescribed and monitored with great care, and the potential risks must be judiciously weighed against the potential benefits for every client on an individual basis.

References

BALLENGER, J. (1986). Pharmacotherapy of the panic disorders. *Journal of Clinical Psychiatry, 47* (Supplement), 27–31.

BIEDERMAN, J. (1988). Pharmacological treatment of adolescents with affective disorders and attention deficit disorder. *Psychopharmacology Bulletin, 24,* 81–87.

CANTWELL, D. (1985). Pharmacotherapy of ADD in adolescents: What do we know, where should we go, how should we do it? *Psychopharmacology Bulletin, 21,* 251–257.

COWDRY, R. (1987). Psychopharmacology of borderline personality disorder: A review. *Journal of Clinical Psychiatry, 48* (Supplement), 15–22.

DAVIS, G. (1985). Psychopharmacology. In H. Sudak (Ed.), *Clinical psychiatry* (pp. 435–452). St. Louis: Warren Green.

DELVA, N., & LETEMENDIA, F. (1982). Lithium treatment in schizophrenia and schizoaffective disorder. *British Journal of Psychiatry, 141,* 387–400.

DeVEAUGH-GEISS, J., LANDAU, J., & KATZ, R. (1989). Preliminary results from a multi–center trial of clomipramine in obsessive-compulsive disorder. *Psychopharmacology Bulletin, 25,* 36–40.

EXTEIN, I., & GOLD, M. (1988). The treatment of cocaine addicts: Bromocriptine or desipramine. *Psychiatric Annals, 18,* 535–537.

FAWCETT, J. (1989). Valproate use in acute mania and bipolar disorders: An international perspective. *Journal of Clinical Psychiatry, 50* (Supplement), 10–12.

FRANCE, R., HOUPT, J., & ELLINWOOD, E. (1984). Therapeutic effects of antidepressants in chronic pain. *General Hospital Psychiatry, 6,* 55–63.

GABBARD, G. (1990). *Psychodynamic psychiatry in clinical practice.* Washington, DC: American Psychiatric Press.

GITLIN, M. (1990). *The psychotherapist's guide to psychopharmacology.* New York: Macmillan.

GOLDBLOOM, D., & OLMSTEAD, M. (1993). Pharmacotherapy of bulimia nervosa with fluoxetine: Assessment of clinically significant attitudinal change. *American Journal of Psychiatry, 150*(5), 770–774.

HENINGER, G., CHARNEY, D., & STERNBERG, D. (1983). Lithium carbonate augmentation of anti-depressant medications: An effective prescription for treatment of refactory depression. *Archives of General Psychiatry, 40,* 1335–1342.

KALES, A., SOLDATOS, C., BIXTER, E., & KALES, J. (1983). Early morning insomnia with rapidly eliminated benzodiazepines. *Science, 220,* 95–97.

KANE, J., QUITKIN, F., RIFKIN, A., RAMOS-LORENZI, J., NAYAK, D., & HOWARD, A. (1982). Lithium carbonate and imipramine in the prophylaxis of unipolar and bipolar II illness: A prospective placebo-controlled comparison. *Archives of General Psychiatry, 39,* 1065–1069.

LUBORSKY, L., SINGER, B., & LUBORSKY, L. (1975). Comparative studies of psychotherapies. *Archives of General Psychiatry, 32,* 602–611.

NOYES, R. (1985). Beta-adrenergic blocking drugs in anxiety and stress. *Psychiatric Clinics of North America, 8,* 119–132.

RAFT, D., DAVIDSON, J., MATTOX, A., MUELLER, R., & WASIK, J. (1979). Double-blind evaluation of phenelzine, amitriptyline, and placebo in depression associated with pain. In A. Singer (Ed.), *Monoamine oxidase: Structure, function, and altered functions.* New York: Academic Press.

RASKIND, M., RISSE, S., & LAMPE, T. (1987). Dementia and antipsychotic drugs. *Journal of Clinical Psychiatry, 48* (Supplement), 16–18.

SALZMAN, B. (1988). Use of benzodiazepines to control disruptive behavior in inpatients. *Journal of Clinical Psychiatry, 49* (Supplement), 13–15.

SALZMAN, B. (1991). *The handbook of psychiatric drugs.* New York: Henry Holt.

SCHATZBERG, A., & COLE, J. (1991). *Manual of clinical pharmacology* (2nd ed.). Washington, DC: American Psychiatric Press.

SMITH, M., GLASS, G., & MILLER, T. (1980). *The benefits of psychotherapy.* Baltimore: Johns Hopkins University Press.

YUDOFSKY, S., HALES, R., & FERGUSON, T. (1991). *Everything you need to know about psychiatric drugs.* Washington, DC: American Psychiatric Press.

Bibliography

BEITMAN, B., & KLERMAN, G. (1991). *Integrating pharmacotherapy and psychotherapy.* Washington, DC: American Psychiatric Press.

CIRAULO, D., SHADER, R., GREENBLATT, D., & CREELMAN, W. (Eds.). (1989). *Drug interactions in psychiatry.* Baltimore: Williams and Wilkins.

DUBOVSKY, S. (1988). *Concise guide to clinical psychiatry.* Washington, DC: American Psychiatric Press.

DULCAN, M., & POPPER, C. (1991). *Concise guide to child and adolescent psychiatry.* Washington, DC: American Psychiatric Press.

GABBARD, G. (1992). Psychodynamic psychiatry in the decade of the brain. *American Journal of Psychiatry, 149*(8), 991–998.

KANE, J., & LIEBERMAN, J. (1992). *Adverse effects of psychotropic drugs.* New York: Guilford.

MAXMEN, J. (1991). *Psychotropic drugs: Fast facts.* New York: Norton.

SELIGMAN, L. (1990). *Selecting effective treatments.* San Francisco: Jossey-Bass.

CHAPTER EIGHT

PRACTICAL ISSUES WITH CLIENTS

You will most likely be faced with a variety of potentially challenging counseling experiences and anxiety-provoking clinical issues during your internship, which may leave you feeling unsettled and wondering just how to proceed. This chapter presents practical, basic guidelines for addressing some of those difficult client problems and treatment situations that you may be encountering for the first time as a counselor intern. However, we strongly encourage you to use additional resources whenever you have questions concerning your interactions with clients. Discuss complex situations or dilemmas with your supervisor or other clinicians at your internship site; ask your counseling program professors for help; read professional books or journal articles that provide current clinical information; refer to the Ethical Standards of the American Counseling Association (see Appendix E); review policies and procedures outlined by your agency.

The particular needs and problems of your client, as well as the demands and constraints of the immediate clinical circumstances, will dictate your counseling style, treatment strategies and goals, and specific therapeutic interventions. The guiding principle in every case, however, involves working toward engagement of the client in a strong therapeutic alliance. This therapeutic alliance enables you to collect data for assessing the client, to develop a compassionate and evolving understanding of the client, to formulate an appropriate treatment plan, and to begin helping (Shea, 1988). We have identified seven general goals that provide a common platform for the engagement process and are well-suited for all counseling situations:

1. Provide a safe space, physically and psychologically
2. Establish rapport
3. Employ a collaborative stance
4. Understand your client's concerns contextually
5. Instill hope

6. Identify and mobilize your client's internal and external resources and support systems
7. Identify and mobilize your own resources and support systems*

The following question-and-answer sections contain specific information and practical recommendations for managing some of the perplexing, problematic, complicated, or possibly stressful treatment situations you may experience during your counseling internship. We have also provided suggested readings wherever possible, so that you may explore certain areas in more depth.

VIOLENCE

My next client appears angry and agitated and has been arguing loudly with several individuals in the waiting area. The referral history reveals that the client picked up a chair and threw it across the room last month during a disagreement with his brother. How should I proceed with this client?

Unfortunately, aggressive, acting-out behaviors occur more often within the context of clinical interactions than we would like to think. Statistics indicate that 17% of emergency room patients are violent and that 40% of psychiatrists are physically assaulted at least once during their careers (Tardiff, cited in Shea, 1988). Safety should be your first consideration with potentially violent clients. As a counselor intern, keeping safe includes

1. being alert and recognizing potentially threatening clients and/or dangerous situations
2. being aware of the ways your own behavior may either escalate or defuse a client's anger
3. taking precautions to protect yourself against physical assault

Violence may be a possibility when you are dealing with clients who

1. are alcohol intoxicated or who have ingested drugs
2. have disorders that include psychotic processes or organic brain dysfunctions
3. have underlying paranoid ideation
4. are being restrained or committed involuntarily
5. have poor impulse control
6. are in the midst of family feuding, with other family members present*

Shea (1988) noted that certain warning signs should alert the clinician to impending dangerous behavior on the part of the client. He included the following:

*From *Psychiatric Interviewing: The Art of Understanding*, by S. C. Shea. Copyright © 1988 by W. B. Saunders Company. Reprinted by permission.

- The client's verbalizations become increasingly angry, sarcastic, challenging, or threatening.
- The client begins pacing, assumes a boxer-like stance, or won't remain seated.
- The client begins vigorous finger-pointing, shakes clenched fists, raises a hand or fist over the head, pounds a fist into the other palm, or pantomimes choking someone.
- The client grasps the arms of the chair or other objects very tightly, so that knuckles turn white, while leaning forward in the chair.
- The client snarls or bares teeth.*

To calm the agitated client and to help prevent violence, you may adhere to the following behavioral guidelines:

- Position yourself in front of the client.
- Speak slowly in a normal conversational tone.
- Carefully explain all your actions in order to avoid arousing suspicion (such as, "I'm going to reach across this table to pick up the pen now").
- Avoid too much eye contact.
- Do not touch the client.
- Provide extra interpersonal space.
- Encourage the client to verbalize feelings rather than acting them out.
- Make sure that the client is able to save face.
- Do not challenge the truthfulness of any of the client's statements (Gabbard, 1990; Shea, 1988).

If you find yourself about to face (1) any new client about whom you have little information, (2) a potentially aggressive client, or (3) a situation that could become dangerous, take steps to ensure your safety. Make certain that help is available and that you will be able to summon assistance quickly and easily. Seat yourself closer to the door; do not allow your client to place himself or herself between you and the door at any time. You may keep the door to the room ajar or open during the session if this seems appropriate. Do not turn your back to the client. That means the client should enter the room ahead of you at the start of the session. At the end of the session, you should exit first, but keep your body turned partially to the client so that you can observe his or her behavior at all times. Trust your intuition. If you feel that you are in danger, explain to your client that you are uncomfortable and need to leave. Notify other staff members immediately, so that the client is not left unattended. If a client produces a weapon, leave the room *immediately*, even if the client is not threatening you personally, and summon help. One counselor intern was cut when he got too close to a client with borderline personality disorder who pulled out a knife and began slashing her wrists.

---

*From *Psychiatric Interviewing: The Art of Understanding*, by S. C. Shea. Copyright © 1988 by W. B. Saunders Company. Reprinted by permission.

Suggested Readings on Violence

BRIZER, D., & CROWNER, M. (Eds.). (1989). *Current approaches to the prediction of violence.* Washington, DC: American Psychiatric Press.
DICKSTEIN, L., & NADELSON, C. (Eds.). (1988). *Family violence: Emerging issues of a national crisis.* Washington, DC: American Psychiatric Press.
DURAND, V. (1991). *Severe behavior problems.* New York: Guilford.
PETERS, R., McMAHON, R., & QUINSEY, V. (Eds.). (1992). *Aggression and violence throughout the lifespan.* Newbury Park, CA: Sage.
ROTH, L. (1987). *Clinical treatment of the violent person.* New York: Guilford.
STREAN, H. (Ed.). (1984). *Psychoanalytic approaches with the hostile and violent patient.* Binghamton, NY: Haworth.
TARDIFF, K. (1989). *Concise guide to assessment and management of violent patients.* Washington, DC: American Psychiatric Press.
TOTH, H. (1992). *Violent men.* Washington, DC: American Psychological Association.

SUICIDAL BEHAVIOR

My client appears to be very depressed and has made the comment, "I really think I'd be better off dead." How do I decide what to do next?

All suicidal statements need to be taken seriously and thoroughly evaluated. About two-thirds of those people who kill themselves first inform a family member, friend, or physician of their plans, and suicide occurs in 15% of cases of depression (Dubovsky, 1988). Suicidal feelings and impulses are so prevalent that one study reported fully 10% of high school students between the ages of 16 and 19 who were not previously identified as upset had seriously considered killing themselves during the previous week (Farberow, Litman, & Nelson, cited in Farberow, Heilig, & Parad, 1990). Some interns are afraid that if they ask a question such as "Have you been thinking about hurting or killing yourself?" they will be giving the client the idea. Asking about suicidal thoughts will *not* encourage your client to commit suicide. As a counselor intern, you should attempt to elicit suicidal ideation with every client, and you should certainly explore and assess all suicidal thoughts or statements to determine lethality, so that you can take action to help ensure your client's safety.

The client who verbalizes suicidal wishes is really expressing the extent of his or her hopeless, helpless, lonely, or otherwise painful feelings. You can first respond with empathy to your client's distress, and then you can proceed to evaluate risk factors. Ask about intensity of suicidal feelings, as well as frequency and duration of suicidal thoughts ("How often do you feel as though you want to die?" or "How long did you stand on the bridge imagining that you were dead?"). You should also ask whether the client has a concrete plan ("Have you been thinking about how you would take your life? Tell me about what you would do.") or the means to carry out this plan ("Do you have access to a gun?" or "How many of those pills left over from your prescription do you have at home right now?''). Clients who present suicidal ideation with a plan and the means to execute the plan usually present a higher risk. When assessing clients for suicidal intent, you may wish to use a clinical instrument, such as the SAD PERSONS scale (Patterson, Dohn, Bird, & Patterson, 1983). This scale is an acronym that helps mental health professionals weigh pertinent suicidal risk factors. High-risk factors include

<u>S</u>ex (males)
<u>A</u>ge (older clients)
<u>D</u>epression
<u>P</u>revious attempt
<u>E</u>thanol (alcohol) abuse
<u>R</u>ational thinking loss
<u>S</u>ocial support system lacking (lonely, isolated clients)
<u>O</u>rganized plan
<u>N</u>o spouse
<u>S</u>ickness (particularly chronic or terminal illness)

We would also like to note several other situations that should alert you to increased suicidal potential in your client:

1. The client has been taking psychotropic medications to combat severe depression for several weeks and has regained some energy, so that, paradoxically, he or she now feels ''improved enough'' to be able to carry out plans to kill himself or herself.
2. Family members or friends have committed suicide. Anniversary dates of those events may be especially difficult.
3. The depressed client appears suddenly calm or speaks of feeling relieved, of having found a ''solution'' to his or her pain. Always ask about the source of relief.
4. The client gives away possessions or makes arrangements concerning legal or financial issues. One client explained to a counselor intern that she had visited the beauty shop, the bank, her attorney, and then had made a trip to the funeral parlor to pick out a coffin on the day before she intended to shoot herself. Fortunately, she communicated her intent to the counselor intern, as well.
5. Your intuition tells you that the client is not doing well or is upset enough to hurt himself or herself.

Once you have identified, explored, and assessed suicidal ideation and determined that there is some significant risk, what can you do next? First, as a counselor intern, be certain to let your supervisor know about the situation before you allow the client to leave. Depending on the circumstances and your experience, your supervisor may want to interview the client or to review your plans. Next, mobilize the client's own resources, such as religious values and beliefs, family, friends, and the healthy part of the self that is seeking help. Educate the client concerning hopelessness, because depressive disorders and suicidal feelings are often time limited and treatable. Secure a signed behavioral contract in which the client agrees not to hurt himself or herself until the next counseling session and/or to telephone for help if suicidal feelings become unmanageable. Provide a 24-hour telephone number for your client, in accordance with agency regulations, such as a crisis hot-line, psychiatric emergency service, or clinician on call. Maintain a close, supportive relationship with the client and increase the frequency of your counseling sessions, even seeing the client daily if necessary, until the crisis subsides. If the client cannot agree to keep herself or himself safe or if your gut feeling tells you that the client is still in trouble, use other resources. Consult with your supervisor or other clinicians on site. If hospitalization is indicated, follow agency procedures. Try to encourage voluntary hospitalization for the seriously suicidal client by discussing the need for "getting all the help we possibly can to keep you safe." Involuntary hospitalizations may unfortunately sometimes be necessary, and in these cases you should follow agency, county/state, and hospital policies and regulations carefully. In all instances, carefully document the client's mood, affect, behavior, and exact statements, as well as your interventions and plans.

Suggested Readings on Suicidal Behavior

BERMAN, A., & JOBES, D. (1991). *Adolescent suicide: Assessment and intervention.* Washington, DC: American Psychological Association.
BONGAR, B. (1991). *The suicidal patient: Clinical and legal standards of care.* Washington, DC: American Psychological Association.
JACOBS, D. (Ed.). (1992). *Suicide and clinical practice.* Washington, DC: American Psychiatric Press.
LANN, I., MOSCICKI, E., & MARIS, R. (Eds.). (1989). *Strategies for studying suicide and suicidal behavior.* New York: Guilford.
LEENAGERS, A., MARIS, R., McINTOSH, J., & RICHMAN, J. (1992). *Suicide and the older adult.* New York: Guilford.
WHITAKER, L., & SLIMAK, R. (Eds.). (1990). *College student suicide.* New York: Haworth.

CHILDREN

> One of my clients is an eight-year-old child who is not very verbal and who appears to be quite anxious. How can I help this youngster?

Counseling children often requires that you modify your professional role, your style, and your treatment modalities. You may need to be cognizant of the impact on your client of various "systems," such as the family, the school, the neighborhood group of children, and sometimes the court and/or other social service agencies. Children are dependent on the adults around them, and therefore you may find that you need to act as an advocate for your client and as a liaison to all these powerful outside forces. Most of the time, the child client is brought to counseling or therapy because some adult in his or her life believes there is a problem, not because the child wants to be there. Your first task with children is to explain the treatment situation, your relationship, and your tentative goals.

If you work with children during your internship, you need to be well-grounded in developmental theory, so that you can assess your client's cognitive, emotional, social, behavioral, and physical development. A thorough understanding of where the child stands developmentally allows you to

1. relate to the child most effectively to establish trust and a therapeutic alliance
2. determine whether problems or symptoms represent significant deviations from normal development
3. identify goals for change
4. formulate treatment plans in specific areas, to help children "get back on track" developmentally

Children's natural tendency to adapt and to grow, in addition to their inherent resiliency, helps clinicians to "adopt the goal of returning children to healthy developmental pathways as an organizing schema for child psychotherapy" (Vernberg, Routh, & Koocher, 1992, p. 73).

Children often express themselves more easily through art and play therapy than through the more traditional "talking therapy." Even very verbal, bright children may not possess the vocabulary or linguistic ability to describe their perceptions, affective experiences, attitudes, or difficulties (Hughes & Baker, 1990). The child can more easily "play out" or draw powerful feelings, upsetting thoughts, worrisome situations, or past events. Play or art are helpful treatment modalities because they

1. engage the child immediately by providing an enjoyable, nonthreatening activity and attractive materials
2. afford the child emotional release (catharsis)
3. separate painful or troubling images from the child's self, so they are not so secret any more
4. increase the bond between yourself and the child as you share these feelings and experiences
5. empower the child by giving him or her some symbolic control over situations where, in reality, he or she has little or no control

For play therapy, we recommend simple toys, such as a family of dolls, stuffed animals, hand puppets, blocks, cars and trucks, and perhaps a tea set, doctor set, or basic board game. Art materials should also be simple and geared to the child's developmental level, as well as to your own tolerance for messiness. You may select crayons, markers, play-dough, collage materials, or various types of paint. Some art materials are more easily controlled than others, such as crayons as compared to water-color paint, and this is an important consideration, also.

To understand your child client's play and art, you will need to remember to think metaphorically (Dulcan & Popper, 1991). The play or art is a symbolic representation of the child's inner self and his or her perceptions of the world. We recommend that you keep your interpretations and interventions within the metaphor that the child has created. Do not connect your interventions too closely to the child, because the child will feel threatened and communication will be hampered. For example, you may ask your child client, "What is this little girl doll feeling like when the mommy doll goes away?" instead of, "Did you feel sad like this dolly does when your mommy had to go to the hospital?" You may say, "I see that you drew a man shooting a great big gun. Tell me about what he is going to do next," instead of "Are you feeling angry enough to want to shoot someone?"

Recent studies (Lambert, Shapiro, & Bergin, cited in Rubin & Niemeier, 1993) indicate that cognitive and behavioral approaches provide superior outcomes when dealing with behavioral problems in children. Cognitive-behavioral techniques especially suited for counseling children include establishing small, attainable behavioral goals within a manageable time frame and providing positive reinforcement. For example, the child will remember to bring all homework assignments home from school for three consecutive days and will be rewarded with praise each day that the goal is accomplished and with a small treat after three days of success. Goals can be increased and time periods can be gradually extended as the child achieves success. Parents and teachers need to be closely involved, so that the child experiences consistent support and guidance in developing self-control.

Working successfully with a child client involves being creative and flexible in tailoring your therapeutic approach to your client's developmental level, as well as being sensitive to the child's unique personality and needs. Patience, warmth, a sense of humor, and an honest enjoyment of children are also helpful characteristics of the child therapist.

Suggested Readings on Children

AXLINE, V. (1969). *Play therapy*. New York: Random House.

BRASWELL, L., & BLOOMQUIST, M. (1991). *Cognitive-behavioral therapy with ADHD children*. New York: Guilford.

BROMFIELD, R. (1992). *Playing for real: The world of a child therapist*. New York: Dutton.

DULCAN, M., & POPPER, C. (1991). *Concise guide to child and adolescent psychiatry*. Washington, DC: American Psychiatric Press.

GIL, E. (1991). *The healing power of play*. New York: Guilford.

HUGHES, J., & BAKER, D. (1990). *The clinical child interview*. New York: Guilford.

KENDALL, P., & BRASWELL, L. (1993). *Cognitive-behavioral therapy for impulsive children*. New York: Guilford.

KLEPSCH, M., & LOGIE, L. (1982). *Children draw and tell*. New York: Brunner/Mazel.

SCHAEFER, C., & O'CONNOR, K. (Eds.). (1983). *Handbook of play therapy*. New York: Wiley.

THOMPSON, C., & RUDOLPH, L. (1992). *Counseling children* (3rd ed.). Pacific Grove, CA: Brooks/Cole.

CLIENT RESISTANCE

My client has been arriving late for our sessions or sometimes skipping her appointments altogether. She often "jokes around" rather than talking about any of her issues when she does come for counseling. I know this client is being resistant, but I'm not sure why, or what to do about it.

Corey (1991) defines resistance as "anything that works against the progress of therapy. Resistance refers to any idea, attitude, feeling, or action (conscious or unconscious) that fosters the status quo and gets in the way of change" (p. 122). Change is generally uncomfortable. Resistance comes into play when the client tries to avoid dealing with painful, threatening, or unpleasant thoughts and feelings. You can therefore understand resistance as a defense against anxiety and as a protective coping mechanism, albeit a maladaptive one in counseling. As a novice counselor, you will need to remember that your client is not simply being "uncooperative" and that the resistant behavior is not meant to frustrate you (Corey, 1991; Kahn, 1991). The resistance needs to be appreciated as a meaningful, often unconscious, defensive dynamic, which can serve to help you better understand your client's inner world. Shea (1988) explained:

A seasoned clinician eventually recognizes resistance not as a demon but as an odd ally of sorts, offering an opportunity for insight. The presence of resistance serves to alert the clinician that the anxieties and defenses of the patient are near the surface. If attended to sensitively, resistance is a pathway to understanding. (p. 500)

One strategy in addressing your client's resistance is encouraging your client to explore the source of the resistant behaviors. When your client becomes aware of the meaning and need for such behaviors as being late, skipping sessions, remaining silent for long periods of time, asking questions about your personal life, telling a great many jokes, and so forth, he or she will be more ready to acknowledge and master the underlying painful issues that he or she has been avoiding. Often, just your empathic comment may help your client overcome resistance to discussing difficult issues: "It's hard for you to get started talking about your father, isn't it?" Another strategy in dealing with a resistant client involves going "with" the resistance, rather than fighting "against" it (Corey & Corey, 1992; Shea, 1988). Moving "with" the resistance entails flexibility and creativity on your part. For example, Corey and Corey (1992) suggested inviting a reluctant adolescent to join a group for a three-session trial period, rather than forcing him or her to participate in the group.

Suggested Readings on Client Resistance

ARONSON, M., & SCHARFMAN, M. (1992). *Psychotherapy: The analytic approach.* Northvale, NJ: Jason Aronson.

BENJAMIN, A. (1987). *The helping interview.* Boston: Houghton Mifflin.

GABBARD, G. (1990). *Psychodynamic psychiatry in clinical practice.* Washington, DC: American Psychiatric Press.

KAHN, M. (1991). *Between therapist and client.* New York: Freeman.

McCOWN, W., & JOHNSON, J. (1992). *Therapy with treatment resistant families.* New York: Haworth.

SHEA, S. (1988). *Psychiatric interviewing: The art of understanding.* Philadelphia: Saunders.

TRAVERS, J., & STERN, M. (Eds.). (1985). *Psychotherapy and the uncommitted patient.* Binghamton, NY: Haworth.

URSANO, R., SONNENBERG, S., & LAZAR, S. (1991). *Concise guide to psychodynamic psychotherapy.* Washington, DC: American Psychiatric Press.

CLIENT REFERRALS

How do I refer a client who will need to continue treatment after leaving my agency?

Most of the time, your agency will outline referral procedures and will have a list of community resources available. If your agency does not, we suggest that you begin assembling your own network of outside therapists and other resources as soon as you can. Your supervisor may be able to offer you suggestions concerning referrals and after-care planning for your client. You can also consult with other clinicians at your site, as well as your counseling program professors. Very often, various human service agencies that do not provide the particular service your client needs will offer names and telephone numbers of other places that do. Collect names of professionals you meet at workshops or meetings; ask colleagues to share names of any good therapists they have encountered; call colleges and universities in your area that offer graduate programs in counseling, social work, or psychology to ask for referral sources. If your client has a specialized problem related to health or aging, you may call local hospitals and request referral sources from gerontological services, rehabilitation medicine, psychiatry departments, and so on. (You're especially fortunate if you have a teaching hospital or university medical center in your area.)

To begin the referral procedure, explore your client's needs and preferences in detail and try to accommodate your client as much as possible so that he or she will be more likely to follow through on the after-care plans. For example, consider the following questions:

- Does your client require a therapist (for continued work on emotional issues), a marriage counselor (for marital/relationship problems), a case manager (to coordinate services and to provide help with transportation, housing, etc.), a psychiatrist (to prescribe or monitor medications or to manage serious psychiatric disorders)?

- Is your client able to drive, and if so, how far a distance would your client be willing to go for appointments?
- Does your client have a preference for a male or female clinician, or an older or younger professional?
- Do you feel that your client would relate better to a more supportive individual or perhaps to someone who is more directive?
- Does your client require evening or weekend appointments rather than weekday times?

You will not be able to find the ''perfect'' helper for your client, but you can try to meet as many criteria as seems feasible.

The next step in your referral procedure involves choosing a few resources that seem to be good possibilities and obtaining signed releases of information from your client so that you can telephone or meet with these professionals to discuss your client's needs and problems. Most clients will be very appreciative of your efforts to help them in this way. Call or arrange to speak with the potential referral sources to gather a bit more information so that you and your client can make a final decision together. When you do call these clinicians, keep in mind how quickly they call you back, whether they speak with you themselves or have their secretaries answer your questions, and whether they appear to be patient, interested, warm, and experienced with your client's particular problems. Ask how soon they will be able to give your client an appointment and what kinds of insurance they accept. Make certain to thank all the professionals for their time and help, and let them know that you will think of them for future referrals.

Next, share pertinent information concerning the best two or three referral sources with your client, but encourage your client to make the final selection. Explain to your client that he or she may want to try out more than one of these helpers to find the best fit. You may need to offer support for some clients and to actually be there, standing by, as they telephone to set up their first appointments. Finally, complete necessary paperwork and sign releases of information forms so that you can send client records to the clinician who will provide aftercare.

SEXUAL ABUSE

I am aware that sexual abuse is a serious issue. How can I recognize symptoms of sexual abuse in children? What should I do if I suspect that a child client has been sexually abused? How should I deal with the adult client who was sexually abused in childhood?

Statistics today indicate that one out of every three adult females has been sexually abused as a child (Russell, 1984) and that between 10% and 20% of adult males have also suffered childhood sexual abuse (Briere, 1992). In addition, the continuing sexual abuse of children is reported almost daily in the media. As a mental health professional, you need to be cognizant of the signs and symptoms of current abuse and aware of the emotional sequelae of past abuse. You need to know how to go about reporting cases of child abuse, and you also need to be sensitive to the issues concerning the treatment of sexually abused children, as well as the treatment of adult survivors of sexual abuse.

Eliana Gil (1988) wrote that the following internalizing behaviors are indicators of abuse in children, but she cautioned that the presence of these behaviors does not constitute conclusive proof of abuse. According to Gil (1988), abused children

- appear withdrawn and unmotivated to seek interactions
- exhibit clinical signs of depression
- lack spontaneity and playfulness
- are overcompliant
- develop phobias with unspecified precipitants
- appear hypervigilant and anxious
- experience sleep disorders or night terrors
- demonstrate regressed behavior
- have somatic complaints (headaches, stomachaches)
- develop eating disorders
- engage in substance and drug abuse
- make suicide gestures
- engage in self-mutilation
- dissociate (pp. 9–10)*

Gil (1988) also described several externalizing behaviors that are often evident in abused children. She wrote that such children are

- aggressive, hostile, and destructive
- provocative (eliciting abuse)
- violent, and may kill or torture animals
- sexualized (p. 10)*

Child abuse must be reported, so that steps may be taken to intervene and protect the child. If you suspect that a child has been abused, or if you learn about a case of child abuse from a client, consult with your site supervisor and counseling program supervisor immediately. Your agency may have specific guidelines concerning the reporting of child abuse. You should follow all agency, county, and state procedures carefully, and document everything in detail. When you make the telephone call to report the abuse, you will

*From *Treatment of Adult Survivors of Childhood Abuse,* by E. Gil. Copyright ©1985 by Eliana Gil. Reprinted by permission of Launch Press.

need to provide such information as the name, age, and address of the child; the name, age, and address of the abuser (if known); the name and address of the child's parent or guardian (if different from the abuser); the exact nature of the abuse; the location and date where the abuse occurred. You will also be asked your name, position, and relationship to the child or the informant.

Reporting child sexual abuse is difficult for counselor interns, at best; one intern described her experience as "traumatic." Very often, the abuser is a family member and the child has mixed feelings for, or even loves, this individual and does not want to see him or her "get in trouble." The other family members may be upset and angry at the child and at you, too. The possibility that the child will be removed from the home or that the family will be destroyed is very real, very frightening for the child, and very disconcerting for the counselor intern, as well. If you find, during your internship, that you must report an incident of child abuse, use your site supervisor, your counseling program supervisor, and other professors from your counseling department for support. Sharing your feelings with other students in your internship discussion group may also prove to be helpful.

In counseling a child victim of sexual abuse, you will first need to work gently and patiently to build a trusting alliance. A consistent, reliable therapeutic framework and respectful attention to boundaries are essential. Children who have been abused have often had their capacity to trust shattered; these children have been betrayed and hurt by the very adults on whom they were dependent for love, care, and protection. In the case of incest, the child will believe that he or she is the "bad one" who either caused or deserved the abuse, because the cognitive schema of young children categorizes parents or other adult family members as good, smart, right, all-knowing, and all-powerful. The child avoids cognitive dissonance by refusing to acknowledge the parent as the source of wrongdoing. Treatment goals in counseling sexually abused children include addressing the following ten issues (Porter, Blick, & Sgroi, 1982):

1. "Damaged Goods" syndrome
2. Guilt
3. Fear
4. Depression
5. Low self-esteem and poor social skills
6. Repressed anger and hostility
7. Impaired ability to trust
8. Blurred role boundaries and role confusion
9. Pseudomaturity coupled with failure to accomplish developmental tasks
10. Self-mastery and control (p. 109)*

*Reprinted with the permission of Lexington Books, an imprint of Macmillan, Inc., from *Handbook of Clinical Intervention in Child Sexual Abuse* by Susanne M. Sgroi. Copyright © 1982 by Lexington Books.

Childhood sexual abuse has serious, long-term consequences and affects the adult survivor in many realms. The following list of psychological sequelae to past sexual abuse is based on the work and ideas of John Briere (1992):

- *Post-traumatic stress disorder* (hypervigilance, exaggerated startle response, intrusive thoughts and efforts to avoid these thoughts, frightening images and flashbacks, nightmares, etc.)
- *Cognitive distortions* (low self-esteem, guilt, shame, imagined danger, feelings of powerlessness, etc.)
- *Dysphoric moods* (major depression, dysthymia, anxiety, uncontrollable anger)
- *Dissociation* ("spacing out," numbing, amnesia, feelings of unreality or not belonging to the human race, depersonalization, ego states, multiple personality disorder)
- *Impaired sense of self* (identity confusion, problems with boundaries, absence of self-soothing abilities, inability to recognize or experience own internal or affective states, lack of awareness of own needs)
- *Disturbed interpersonal connections* (extreme dependency, avoidance of intimacy, sexual dysfunction, promiscuity, aggression)
- *Avoidance behaviors* (chemical dependency, self-mutilation, addictions or compulsions such as exercise or shopping or sex, eating disorders, suicidal actions)*

During your internship you are quite likely to counsel an adult survivor of past sexual abuse because of the ubiquity of the problem. Trust is an important issue with adults, just as it is with children. Be reliable and dependable; expect your client to question the safety of your relationship many times. You can begin the counseling process by believing and validating your client. Many children who tried to "tell" about their abuse were punished for "lying." Encourage your client to tell his or her story now, in adulthood, so that the secret loses its power. You will need to work extensively on issues of shame, guilt, and anger. Group therapy is often very helpful and serves to lessen isolation, shame, and hopelessness. Educate your client about the process of recovery (after you have done some reading in this area!) and reassure your client during the phase when he or she is recalling painful memories. You may also offer concrete suggestions for dealing with flashbacks and explain that the repressed memories are allowed back into consciousness only when the mind is ready to cope with them.

Working with sexually abused clients can be stressful and emotionally draining, because you are often listening to frightening, unpleasant, and affect-laden material, in addition to watching your client suffer deep pain during a large part of the therapy. Counseling sexually abused clients can be very rewarding, challenging, and interesting, as well, as you help your client move toward health and growth. Counselors who deal extensively with these clients often need their own support group to share concerns and feelings. Call on your site supervisor, program supervisor, or counseling program professors for support if you are feeling overwhelmed.

Suggested Readings on Sexual Abuse

BRAUN, B. (Ed.). (1986). *Treatment of multiple personality disorder.* Washington, DC: American Psychiatric Press.

BRIERE, J. (1992). *Child abuse trauma.* London: Sage.

BROWN, S. (1991). *Counseling victims of violence.* Alexandria, VA: American Counseling Association.

DRAUCKER, C. (1992). *Counselling survivors of childhood sexual abuse.* London: Sage.

GIL, E. (1983). *Outgrowing the pain.* New York: Doubleday.

GIL, E. (1988). *Treatment of adult survivors of childhood abuse.* Walnut Creek, CA: Launch Press.

HERMAN, J. (1992). *Trauma and recovery.* New York: Basic Books.

KLUFT, R. (Ed.). (1990). *Incest-related syndromes of adult psychopathology.* Washington, DC: American Psychiatric Press.

MacFARLAND, K., WATERMAN, J., CONERLY, S., DAMON, L., DURFEE, M., & LONG, S. (1988). *Sexual abuse of young children: Evaluation and treatment.* New York: Guilford.

MALCHIODI, C. (1990). *Breaking the silence: Art therapy with children from violent homes.* New York: Brunner/Mazel.

RENCKEN, R. (1989). *Intervention strategies for sexual abuse.* Alexandria, VA: American Counseling Association.

THE ELDERLY

I am not sure what to do to help one of my clients who is a depressed, elderly man.

Approximately 15% of the elderly population in this country manifest at least moderate emotional problems, but only 2% receive mental health care (Turner & Helms, 1991). Many studies (Dacey & Travers, 1991; Turner & Helms, 1991; Zivian, Larsen, Knox, Gekoski, & Hatchette, 1993) document the disproportionately high rate of psychopathology in persons over 65, as well as the fact that the elderly are grossly underserved by mental health professionals.

Zivian et al. (1993) pointed out that although evidence exists that demonstrates that older adults benefit from counseling and psychotherapy, many mental health professionals are reluctant to treat the elderly. Zivian et al. (1993) attributed this reluctance to

(1) having negative attitudes toward old age (Kimmel, 1988), (2) considering older adults to be inappropriate candidates for psychotherapy (Sparacino, 1978–79), (3) being anxious themselves about growing old and dying (Butler, 1975; Kastenbaum et al., 1972), (4) fearing being associated

with low status clients (Kastenbaum et al., 1978), and finally to (5) the limited training opportunities (Lasoski, 1986; Verwoerdt, 1969), and (6) general paucity of literature in the area of clinical gerontology (Storandt, 1977). (p. 668)

Other factors contributing to the underutilization of mental health services by the elderly may be: (1) the widely accepted notion that depression and other distressing psychological states are a "normal" part of aging and therefore must be tolerated, and (2) the inaccessibility of services to older adults, who may have limited incomes and problems with transportation.

Older adults are often dealing with issues of loss and concomitant feelings of narcissistic injury:

- A decline in physical strength and vigor, sensory acuity (vision, hearing, etc.), youthful appearance, and sexual potency (in males) is related to aging.
- A loss of power, status, and financial security may be associated with retirement from career role.
- A decreased independence and autonomy may occur due to health problems or limited funds.
- Friends, family members, and spouse or significant other may be gone, resulting in decreased social support system.
- Illness associated with aging may cause significant discomfort.

In addition, older adults are confronting their own mortality and the fact that "time is running out." They reminisce about the past and try to reconsider experiences and conflicts in order to derive some sense of meaning and continuity in their lives. Robert Butler (Butler, 1983; Butler & Lewis, 1981) wrote that this process of life review is an integral component in the life cycle. Reviewing life events in an effort to come to terms with the past is "strikingly similar to Erikson's psychosocial stage of integrity versus despair. It may culminate in wisdom, serenity, and peace, or it may produce depression, guilt, and anger" (Turner & Helms, 1991, p. 530).

During your internship, you can be helpful to your elderly clients by assisting them in the life review. You can listen empathically to your clients' stories of the past and understand that this narration is not simply "rambling talk" or "resistance to discussing real issues," but rather a valuable and crucial aspect of the therapy. In addition, you will need to encourage your clients to verbalize feelings and deal with the many issues of loss faced by the elderly. We would like to suggest, also, that a thorough evaluation of your elderly clients by a physician experienced in gerontology is an absolute prerequisite to counseling. A great many cases of depression or impaired thinking are caused by underlying organic disease. A psychiatric consultation is often advantageous, also, because psychotropic medications are useful in alleviating many disturbing psychological symptoms (such as insomnia or paranoia) in older patients.

Suggested Readings on the Elderly

BENGTSON, V., & SCHAIE, K. (Eds.). (1989). *The course of later life: Research and reflections.* New York: Springer.

BIRREN, J., & BENGTSON, V. (Eds.). (1988). *Emergent theories of aging.* New York: Springer.

GEARING, B., JOHNSON, M., & HELLER, T. (Eds.). (1988). *Mental health problems in old age: A reader.* Chichester: Wiley.

LEIBLUM, S., & SEGRAVES, T. (1989). Sex therapy with aging adults. In S. Leiblum & R. Rosen (Eds.), *Principles and practice of sex therapy: Update for the 90's* (pp. 352-381). New York: Guilford.

LITWAK, E. (1985). *Helping the elderly.* New York: Guilford.

MYERS, J. (1989). *Adult children and aging parents.* Alexandria, VA: American Counseling Association.

MYERS, W. (1991). *New techniques in psychotherapy of older patients.* Washington, DC: American Psychiatric Press.

SADAVOY, J., LAZARUS, L., & JARVIK, L. (Eds.). (1991). *Comprehensive review of geriatric psychiatry.* Washington, DC: American Psychiatric Press.

SPAR, J., & LaRUE, A. (1990). *Concise guide to geriatric psychiatry.* Washington, DC: American Psychiatric Press.

CONFIDENTIALITY

How can I explain the limits of confidentiality in our counseling relationship to my adult clients and still expect that they will trust me enough to disclose personal information? How can I explain confidentiality to the young children whom I counsel, so that they will understand?

Our litigious society, rather than solely our own philosophy and personal values or our professional ethical standards, often dictates the parameters of our counseling relationship (Gutheil & Gabbard, 1993). The sanctity of the counselor–client alliance and professional confidentiality are no longer absolute legal standards. Therefore, you must always disclose the limits of confidentiality to your clients at the outset of your counseling relationship. Find out your exact state regulations concerning the situations in which you will need to break confidentiality, and memorize these.

As a novice counselor, you may find the task of defining confidentiality to your clients awkward or threatening. Most clients will be reassured to have this information, however, because many of them will be wondering about the "privacy" of their counseling sessions. You can frame the limits of confidentiality as a need to protect the client and to keep other people safe, as well. You can also stress that strict confidentiality is maintained most of the time, and this is very important to you. We recommend that you try writing out your explanation of confidentiality several different ways, and then practice saying it, by yourself, until you feel most comfortable.

We explain confidentiality to our adult clients this way: "Everything we say to each other during our counseling sessions is absolutely private, except in three instances. By law, I am required to get help for you if you feel like hurting yourself, and I am required to report to the proper authorities if you tell me that you have plans to hurt another person or if I learn of any abuse of children or the elderly. In all of these situations, I would discuss my plans with you first, before talking to anybody else. My only motive would be to protect you and others from harm, to help keep people safe. I'd be glad to answer any questions or hear any comment you may have about what I've just said."

During your internship, and as long as you require supervision, you will also need to discuss your status as a counselor trainee. You may wish to say, "The state requires that all mental health professionals be supervised by more experienced helpers for a number of years to ensure the highest quality of care for all clients. I am currently being supervised by John Doe, who is a Licensed Professional Clinical Counselor (or whatever title the supervisor holds). I will be reviewing your records with him from time to time, so that he can offer suggestions concerning our counseling sessions. However, John Doe will keep everything absolutely confidential."

For children, we modify our explanation of confidentiality a bit: "The things we talk about in here are mostly private, just between us, because it's important that you can tell me anything and everything you are feeling or thinking. But it's really important for me to keep you and other people safe, too. So if you tell me something that makes me worry that you might hurt yourself or somebody else, or if I hear that a child or older person is being hurt, then I would have to get some help. I would have to talk to some other adults who could help us handle the problem. But I would always talk it over with you first."

NONCOMPLIANCE WITH TREATMENT

What can I do about a client who is noncompliant with treatment? One of my clients, a 19-year-old male, was required by his college advisor to attend counseling sessions for three months. He came only one time and has not been back.

Some clients, especially those who are mandated to seek counseling by the court or other institutions, will be noncompliant. You should discuss the situation with your supervisor and adhere to agency guidelines. We recommend that you attempt to contact the client at least three times by telephone and try to encourage him or her to return to counseling. However, if you are unsuccessful in reaching the client or he or she still does not return, then you need to send a letter to the client and to the authorities who mandated treatment detailing the problem and your efforts to resolve it. (Your supervisor will need to sign this letter in addition to you.) Keep a copy of the letter at the agency in the client's chart, and keep another copy for yourself. Also, document everything, including dates and times of telephone calls, in detail in the client's chart.

ADOLESCENTS

I have a 15-year-old client and I'm wondering how to build a therapeutic alliance. What approach would be most helpful when I counsel adolescents?

Adolescence, according to Erik Erikson (1963), is characterized by a search for identity and the consolidation of disparate elements of the self into an integrated whole that is congruent with self-perceptions and evaluation by others. Adolescents are faced with some very difficult developmental tasks as they struggle to find their way from childhood to adulthood. Corey and Corey (1992) described adolescence as a turbulent and possibly lonely period, characterized by many conflicts and paradoxes. During adolescence, boys and girls usually

- relinquish their dependency on the adults in their lives
- begin to separate from their families
- begin to make decisions that will affect their futures
- begin to relate to the opposite sex in new ways
- learn to interact with their peers successfully so that they will belong to a group
- must contend with powerful media and peer-group messages urging them to experiment with alcohol, drugs, and sex
- undergo the rapid physical development and emotional changes of puberty

In addition, adolescents must face the dilemmas, pressures, and stressors of today's fast-paced, increasingly complex, and decreasingly family-oriented society. The result is often teenage depression, eating disorders, alcohol and drug abuse, sexual promiscuity and pregnancy, and juvenile delinquency (Dacey & Travers, 1991; Turner & Helms, 1991). The suicide rate for teenagers has increased over 300% since 1960; 13 adolescents kill themselves each day and teen suicide attempts occur about 300,000 times each year, with guns or poison being the most common means used (Turner & Helms, 1991).

When you counsel adolescent clients, you will need to work hard to establish enough trust to build a therapeutic relationship. You may have more success if you are honest, reliable, and consistent, and if you demonstrate your caring, support, interest, and respect. Group therapy is often the most effective treatment modality for adolescents because of their strong allegiance to their peers (Corey & Corey, 1992; Dulcan & Popper, 1991). The therapy group provides a forum for teenagers to find that they are not alone with their problems, to experience a sense of belonging, to experiment with interpersonal interactions with their peers and the adult group leader(s), to learn to accept and provide honest feedback, and to explore values.

Suggested Readings on Adolescents

BRIGHAM, T. (1988). *Self-management for adolescents*. New York: Guilford.

COREY, G., & COREY, M. (1992). *Groups: Process and practice* (4th ed.). Pacific Grove, CA: Brooks/Cole.

FISHMAN, H. (1988). *Treating troubled adolescents*. New York: Basic Books.

GULOTTA, T., ADAMS, G., & MONTEMAYOR, R. (1993). *Adolescent sexuality*. Newbury Park, CA: Sage.

KENDALL, P. (Ed.). (1991). *Child and adolescent therapy*. New York: Guilford.

KESSLER, J. (1966). *Psychopathology of childhood*. Englewood Cliffs, NJ: Prentice-Hall.

KNOFF, H. (Ed.). (1986). *The assessment of child and adolescent personality*. New York: Guilford.

MORGAN, R. (1990). *Skills for living: Group counseling activities for young adolescents*. Alexandria, VA: American Counseling Association.

ROBIN, A., & FOSTER, S. (1989). *Negotiating parent-adolescent conflict*. New York: Guilford.

SALO, M., & SHUMATE, S. (1993). *Counseling minor clients* (Vol. 4). Alexandria, VA: American Counseling Association.

VERNON, A. (1993). *Developmental assessment with children and adolescents*. Alexandria, VA: American Counseling Association.

HOMICIDAL IDEATION

I am concerned that one of my clients is homicidal. He has made several statements about killing someone, which make me feel uncomfortable.

Homicidal ideation needs to be explored immediately and thoroughly, because there are legal as well as moral and ethical implications if your client actually intends to kill someone. Shea (1988) wrote that "no study has indicated an effective way of accurately predicting homicide. Generally, the factor considered most worrisome is a history of violence, poor impulse control, or attempted homicide" (p. 437). He suggested that a serious potential for homicide exists in situations where there is

- psychosis
- interpersonal conflict
- need for money and other practical concerns
- revenge
- political concerns
- organized crime
- pathological murder for pleasure (p. 440)*

We suggest that if your client verbalizes homicidal ideation during your internship, you have a right to decide whether or not you choose to be the counselor who inquires about the existence of a concrete plan and/or the means

*From *Psychiatric Interviewing: The Art of Understanding*, by S. C. Shea. Copyright © 1988 by W. B. Saunders Company. Reprinted by permission.

to carry out this plan. In either case, you should consult with your site supervisor or other senior clinicians immediately. You do not have enough clinical experience and therefore you should not be assessing the lethality of homicidal ideation during your internship. If you feel too uncomfortable to continue working with the client, discuss the situation and explore alternative treatment options with your supervisor. Make certain to document all interactions and consultations and to let your counseling program supervisor know what is going on also.

SERIOUS PSYCHIATRIC DISORDERS

I am a counselor intern in a psychiatric hospital. How can I help a client who is psychotic, for example, an individual suffering from paranoid schizophrenia?

Generally, clients with active psychotic process have been noncompliant with, or require some adjustment of, their psychotropic medications. Once hospitalized, these people will be placed on a regime of medication and will then require about 10 days to 2 weeks until the blood level of medicine is high enough to help control their psychosis. Until the medication "kicks in" and takes effect, you can be supportive, calm, and very concrete. Do gentle reality testing (for example, "It must be frightening to hear those voices, but they really are part of your illness and your medication is going to help with that") and offer reassurance about the safety of the hospital. Try not to ask too many questions, because many of these patients feel threatened and overwhelmed.

Suggested Readings on Serious Psychiatric Disorders

ANDERSON, C., REISS, D., & HOGARTY, G. (1986). *Schizophrenia and the family.* New York: Guilford.
CHESSICK, R. (1992). *The technique and practice of listening in intensive psychotherapy.* Northvale, NJ: Jason Aronson.
GABBARD, G. (1990). *Psychodynamic psychiatry in clinical practice.* Washington, DC: American Psychiatric Press.
GREDEN, J., & TANDON, R. (1991). *Negative schizophrenic symptoms.* Washington, DC: American Psychiatric Press.
HATFIELD, A. (1990). *Family education in mental illness.* New York: Guilford.
MIRIN, S., GOSSETT, J., & GROB, M. (1991). *Psychiatric treatment.* Washington, DC: American Psychiatric Press.
SEEMAN, M., & GREBEN, S. (1990). *Office treatment of schizophrenia.* Washington, DC: American Psychiatric Press.
SHEA, S. (1988). *Psychiatric interviewing: The art of understanding.* Philadelphia: Saunders.
TALBOT, J. (1987). *A family affair: Helping families cope with mental illness.* Washington, DC: American Psychiatric Press.

TRANSFERENCE/COUNTERTRANSFERENCE

> I have been counseling a client for several weeks and I thought that we had
> a good therapeutic relationship. Recently, however, she has become angry
> at me during our sessions and accuses me of being "rejecting" and "not there
> for her." I actually like this client very much, but now I want to avoid our
> sessions because I find them unpleasant, and I know that I am distancing myself
> emotionally. I'm not sure what is going on or how to get our relationship
> "back on track."

You are experiencing the phenomena known as *transference* and *counter-transference*. These concepts constitute central dynamics in psychoanalytic therapy; however, an appreciation of the importance of transference and countertransference is germane to your understanding of all interpersonal interactions, including all therapeutic relationships.

Transference refers to the *transfer* of feelings, attitudes, fears, wishes, desires, and perceptions that belong to our past relationships onto our current relationships, so that the people in our present lives become the focus of these thoughts and emotions from long ago. Transference occurs for two reasons. First, the human mind constantly compares incoming information with both conscious and unconscious memories in an effort to find some "pattern match," to organize the new information in a meaningful way (Bowers, Regehr, Balthazard, & Parker, 1990; Lewicki, 1985). Therefore, our perceptions of present situations are categorized according to, and linked to, our memories of past situations. Second:

> there seems to be in all people a psychological need to repeat the past,
> in an effort to master that which was difficult or emotionally painful.
> Because psychological development invariably involves difficulty and pain,
> this "compulsion to repeat," and the transferences that result, are ubiquitous
> human experiences. (Ursano, Sonnenberg, & Lazar, 1991, p. 43)

In a counseling relationship, the transference can become especially intense and very real to the client, because the therapy often brings powerful feelings and conflicts from the past into play. In the situation described in the question above, your client may actually be experiencing you as rejecting and emotionally distant, just as some other significant person in her life, probably a parent, used to be. Your client most likely is also experiencing the same hurt feelings, pain, and anger toward you as she did toward that other person, long ago. You can help her by demonstrating that, in reality, you do care very much for her, that you do want to "be there" for her, and that you are trying hard to understand her feelings. You may also explain the concept of transference and encourage her to explore where her feelings may be originating by asking if she has ever felt this way in the past.

Countertransference is similar to transference and refers to the counselor's emotional responses to the client. Countertransference may be due to the therapist's unconscious feelings, thoughts, and conflicts from the past that are evoked by the present clinical situation. However, countertransference can also be interpreted as a reaction to the client's transference, and if understood by the counselor, it can provide insight into the client's emotional state. Ursano et al. (1991) explained that countertransference is experienced by the therapist as a kind of pressure due to the client's transference feelings and can occur in two ways. In the first countertransference reaction, known as concordant countertransference, "the therapist experiences and empathizes with *the patient's emotional position*" (p. 56). In the second, or complementary countertransference reaction, "the therapist experiences and empathizes with the feelings of *an important person from the patient's life*" (p. 56). You seem to be experiencing the second type of countertransference reaction, where you find yourself wanting to avoid sessions because they are unpleasant and you are emotionally distancing yourself from your client. You are experiencing the same rejecting feelings toward your client as another person in her life displayed toward her, long ago. Understanding where your feelings are coming from may help you to get in touch with your real warm and accepting feelings for the client, so that you can reestablish your previous good relationship.

Suggested Readings on Transference/Countertransference

ARONSON, M., & SCHARFMAN, M. (Eds.). (1992). *Psychotherapy: The analytic approach*. Northvale, NJ: Jason Aronson.

CHESSICK, R. (1993). *Dictionary for psychotherapists: Dynamic concepts in psychotherapy*. Northvale, NJ: Jason Aronson.

GABBARD, G. (1990). *Psychodynamic psychiatry in clinical practice*. Washington, DC: American Psychiatric Press.

GIL, M. (1982). *The analysis of transference: Theory and technique* (Vol. 1). New York: International Universities Press.

GOLD, J., & NEMIAH, J. (Eds.). (1993). *Beyond transference*. Washington, DC: American Psychiatric Press.

KAHN, M. (1991). *Between therapist and client*. New York: Freeman.

OGDEN, T. (1992). *The primitive edge of experience*. Northvale, NJ: Jason Aronson.

ROCKLAND, L. (1989). *Supportive therapy*. New York: Basic Books.

URSANO, R., SONNENBERG, S., & LAZAR, S. (1991). *Concise guide to psychodynamic psychotherapy*. Washington, DC: American Psychiatric Press.

GROUP AND INDIVIDUAL THERAPIES

How can I decide whether individual or group therapy would be more helpful for my client?

Individual therapy does not preclude your client's participation in a group; individual and group therapies are often effective concomitant treatment modalities. Corey and Corey (1992) suggested a preferential group format for the adolescent client in the light of (1) the importance of peer inter-actions during the teenage years, and (2) the adolescent's need to separate from adults, which might interfere with the development of a relationship with an individual therapist. Group therapy may also be the more appro-priate treatment modality in instances where individual therapy has been frequently disrupted by intensely negative transferences (for example, in some clients with personality disorders); the presence of other people serves to "dilute" the intensity of the counselor–client relationship somewhat (Yalom, 1985).

Group therapy builds interpersonal skills, provides a sense of belong-ing, and promotes a feeling of universality. Group members come to see that they are not alone with their problems. We identify and discuss the specific healing forces that are often engendered by group process in chapter 4 (Deciding How to Help: Choosing Treatment Modalities). Yalom (1985) wrote that groups are suitable for almost all clients, and he emphasized that the crucial issue is matching the particular client to the right group in terms of the group's focus, structure, and composition. Yalom (1985) explained that the group as a whole, and the other group members individually, may suffer if a new member is added who presents a problem (such as dropping out of the group after only a few sessions).

Yalom (1985) discussed several general types of clients he feels are not suitable candidates for a group therapy. He usually excludes clients who are "brain damaged, paranoid, hypochondriacal, addicted to drugs or alcohol, acutely psychotic, or sociopathic" (p. 228). In addition, Yalom wrote that he attends to more pragmatic concerns by eliminating clients who live too far away geographically from the meeting place and those who must travel fre-quently for business.

During your internship, most of your clients will benefit from a combina-tion of group and individual therapies. Facilitating groups is usually an en-joyable learning experience for counselor interns.

Suggested Readings on Group and Individual Therapies

BEITMAN, B. (1991). *The structure of individual therapy.* New York: Guilford.

BROWN, D., KURPIUS, D., & MORRIS, J. (1988). *Handbook of consultation with individuals and small groups.* Alexandria, VA: American Counseling Associa-tion.

COREY, G. (1991). *Theory and practice of counseling and psychotherapy* (4th ed.). Pacific Grove, CA: Brooks/Cole.

COREY, G., & COREY, M. (1992). *Groups: Process and practice* (4th ed.). Pacific Grove, CA: Brooks/Cole.

LUFT, J. (1984). *Group processes* (3rd ed.). Mountain View, CA: Mayfield.

YALOM, I. (1983). *Inpatient group psychotherapy.* New York: Basic Books.

YALOM, I. (1985). *Theory and practice of group psychotherapy.* New York: Basic Books.

CLIENTS WITH SPECIAL NEEDS

How can I be most helpful to clients who are physically or mentally challenged?
I don't have any training in these areas.

Physical and mental disabilities are tremendous stressors, which may affect every area of a person's life and many of his or her interpersonal relationships, as well. Psychological disorders or emotional upset, therefore, often occurs alongside these physical and mental challenges. You can relate to your clients who have special needs, such as mental retardation, learning disabilities, or physical limitations, in the same way you relate to all your clients: by demonstrating your empathy, your interest, your respect, your nonjudgmental acceptance, your genuineness, your caring, your trustworthiness, and your desire to help. The establishment of a therapeutic alliance provides ''a humanizing social attachment and reciprocal lines of communication'' (Bihm & Leonard, 1993, p. 231), which is healing for all clients, regardless of their special needs.

You will need to be creative and flexible and to ask your client what he or she finds most helpful. An honest approach is best. You can say, for example: ''I haven't counseled anyone before who has multiple sclerosis, but I'm very glad to be working with you. I'm hoping you will feel comfortable enough to tell me just how I can help. I'm going to work hard to understand how you feel and how things are for you.'' We suggest that you consult with your site supervisor and your counseling program supervisor, who may be able to share their insights and experiences with you. You can do some research into your client's special needs at the library, and you can request information from agencies (for example, the Muscular Dystrophy Association) or medical facilities that handle similar problems.

Suggested Readings on Clients with Special Needs

BAUMEISTER, R. (1991). *Meanings of life.* New York: Guilford.

CANTWELL, D., & BAKER, L. (1991). *Psychiatric and developmental disorders in children with communication disorder.* Washington, DC: American Psychiatric Press.

GOODNIK, P., & KLIMAS, N. (1993). *Chronic fatigue and related immune deficiency syndromes.* Washington, DC: American Psychiatric Press.

HANSON, R., & GERBER, K. (1990). *Coping with chronic pain.* New York: Guilford.

MACKLIN, E. (1989). *AIDS and families.* New York: Haworth.

ROURKE, B., & FUERST, D. (1991). *Learning disabilities and psychosocial functioning.* New York: Guilford.

SCHOVER, L., & JENSON, S. (1988). *Sexuality and chronic illness.* New York: Guilford.

SOHLBERG, M., & MATEER, C. (1989). *Introduction to cognitive rehabilitation.* New York: Guilford.

STEWART, D., & STOTLAND, N. (1993). *Psychological aspects of women's health care.* Washington, DC: American Psychiatric Press.

CONCLUDING REMARKS

In this chapter, we have attempted to provide guidelines for handling some of the client problems and treatment situations you may face during your internship. We encourage you to consult with both your site supervisor and your program supervisor and to use other available resources whenever you need help in resolving a challenging problem or a difficult dilemma.

References

BIHM, E., & LEONARD, P. (1993). Counseling persons with mental retardation and psychiatric disorders: A preliminary study of mental health counselors' perceptions. *Journal of Mental Health Counseling, 14*(2), 225–233.

BOWERS, K., REGEHR, G., BALTHAZARD, C., & PARKER, K. (1990). Intuition in the context of discovery. *Cognitive Psychology, 22,* 72–110.

BRIERE, J. (1992). *Child abuse trauma.* London: Sage.

BUTLER, R. (1983). An overview of research on aging and the status of gerontology today. *Milbank Memorial Fund Quarterly: Health and Society, 61*(3), 351–361.

BUTLER, R., & LEWIS, M. (1981). *Aging and mental health.* St. Louis: Mosby.

COREY, G. (1991). *Theory and practice of counseling and psychotherapy.* Pacific Grove, CA: Brooks/Cole.

COREY, G., & COREY, M. (1992). *Groups: Process and practice* (4th ed.). Pacific Grove, CA: Brooks/Cole.

DACEY, J., & TRAVERS, J. (1991). *Human development across the lifespan.* Dubuque, IA: Brown.

DUBOVSKY, S. (1988). *Concise guide to clinical psychiatry.* Washington, DC: American Psychiatric Press.

DULCAN, M., & POPPER, C. (1991). *Concise guide to child and adolescent psychiatry.* Washington, DC: American Psychiatric Press.

ERIKSON, E. (1963). *Childhood and society* (2nd ed.). New York: Norton.

FARBEROW, N., HEILIG, S., & PARAD, H. (1990). The suicide prevention center: Concepts and clinical functions. In H. Parad & L. Parad (Eds.), *Crisis intervention* (Vol. 2, pp. 251–274). Milwaukee, WI: Family Service America.

GABBARD, G. (1990). *Psychodynamic psychiatry in clinical practice.* Washington, DC: American Psychiatric Press.

GIL, E. (1988). *Treatment of adult survivors of childhood abuse.* Walnut Creek, CA: Launch Press.

GUTHEIL, T., & GABBARD, G. (1993). The concept of boundaries in clinical practice: Theoretical and risk-management decisions. *American Journal of Psychiatry, 150*(2), 188–196.

HUGHES, J., & BAKER, D. (1990). *The clinical child interview.* New York: Guilford.

KAHN, M. (1991). *Between therapist and client.* New York: Freeman.

LEWICKI, P. (1985). Nonconscious biasing effects of single instances on subsequent judgments. *Journal of Personality and Social Psychology, 47*(6), 1177–1190.

PATTERSON, W., DOHN, H., BIRD, J., & PATTERSON, G. (1983). Evaluation of suicidal patients: The SAD PERSONS scale. *Psychosomatics, 24,* 343–349.

PORTER, F., BLICK, L., & SGROI, S. (1982). Treatment of the sexually abused child. In S. Sgroi (Ed.), *Handbook of clinical intervention in child sexual abuse* (pp. 109–145). Lexington, MA: Lexington Books.

RUBIN, S., & NIEMEIER, D. (1993). Non-verbal affective communication as a factor in psychotherapy. *Psychotherapy, 29*(4), 596–602.

RUSSELL, D. (1984). *Sexual exploitation: Rape, child sexual abuse, and sexual harassment.* Beverly Hills, CA: Sage.

SHEA, S. (1988). *Psychiatric interviewing: The art of understanding.* Philadelphia: Saunders.

TURNER, J., & HELMS, D. (1991). *Lifespan development* (4th ed.). Chicago: Holt, Rinehart and Winston.

URSANO, R., SONNENBERG, S., & LAZAR, S. (1991). *Concise guide to psychodynamic psychotherapy.* Washington, DC: American Psychiatric Press.

VERNBERG, E., ROUTH, D., & KOOCHER, G. (1992). The future of psychotherapy with children: Developmental psychotherapy. *Psychotherapy, 29*(1), 72–80.

YALOM, I. (1985). *Theory and practice of group psychotherapy.* New York: Basic Books.

ZIVIAN, M., LARSEN, W., KNOX, V., GEKOSKI, W., & HATCHETTE, V. (1993). Psychotherapy for the elderly: Psychotherapists' preferences. *Psychotherapy, 29*(4), 668–674.

Bibliography

ARONSON, A., & SCHARFMAN, M. (1992). *Psychotherapy: The analytic approach.* Northvale, NJ: Jason Aronson.

GITLIN, M. (1990). *The psychotherapist's guide to psychopharmacology.* New York: Macmillan.

HAMILTON, G. (1992). *Self and others: Object relations theory in practice.* Northvale, NJ: Jason Aronson.

JORDAN, J., KAPLAN, A., MILLER, J., STIVER, I., & SURREY, J. (1991). *Women's growth in connection.* New York: Guilford.

LERNER, H. (1988). *Women in therapy.* Northvale, NJ: Jason Aronson.

MADANES, C. (1991). *Strategic family therapy.* San Francisco: Jossey-Bass.

SCHARFF, D. (1992). *Refinding the object and reclaiming the self.* Northvale, NJ: Jason Aronson.

SCHARFF, D., & SCHARFF, J. (1992). *Scharff notes: A primer of object relations theory.* Northvale, NJ: Jason Aronson.

THOMPSON, C., & RUDOLPH, L. (1992). *Counseling children* (3rd ed.). Pacific Grove, CA: Brooks/Cole.

CHAPTER NINE

ALONG THE WAY:
A COUNSELOR SELF-ASSESSMENT

We are in the business of introspection, reflection, analysis, synthesis, and personal action. We personify what Freud (1943) called the "talking cure": our words are our product. It certainly behooves us to look at ourselves as we move through the process of creative self-examination with our clients. Although this self-examination can take considerable courage, it can also help us appreciate how our clients may feel as they expose themselves to us in the therapeutic milieu. We have noticed that, as counselors, we sometimes become desensitized and myopic to the justifiable defenses evident as a client self-discloses issues and concerns, often at our prompting. Our own thorough self-examination may foster increased capacity for empathy and personal and professional refinement. Our own codependency needs may surface in this process, in addition to other possible character quirks (perhaps our own neuroses?). All personal issues are worthy of inspection. For this reason, we recommend personal therapy for all counselors as an adjunct to training. It helps to see the process from the other side.

Dowling (1984) indicated that students possess the ability to accurately evaluate themselves as well as their peers. Further, Bernard and Goodyear (1992) advocated that supervisors teach interns the principles of both peer and self-evaluation and expect such an examination prior to each supervisory session.

Corey and Corey (1993) likewise advised that counselors-in-training do self-exploration so as to better understand their clients' issues. Moreover, they noted that for many students, self-exploration was a more difficult task than that of apprehending their clients' situations. Benjamin (1987), who himself conquered a major obstacle—his blindness—shared his personal philosophy of self-knowledge, which involves trust in our own feelings, ideas, and intuitions. Further, as we become at ease with personal self-exploration, we not only nurture our own personal growth, but also enhance our understanding of clients.

In this chapter, we offer the beginnings of the self-examination process for counselors-in-training. Its purpose is to suggest areas of personal reflection and inner dialogue. Certainly, who we are as people influences choice of profession and how we conduct ourselves as professionals. Thus, we first ask you to look at yourself from a personal vantage point, examining your values, beliefs, needs, priorities, characteristics, and biases. Next, assess your professional role, your career goals, and your personal traits that influence career choice. The chapter is presented in an interrogative format in an attempt to elicit some inner probing. We do, however, advise you to seek out trusted fellow students or professors for pertinent discussion, feedback, and validation.

WHO AM I AS A PERSON?

1. How do I assess my developmental history up to this point of my life? What were the high and low points?
2. When did I realize I was an adult? How did I handle this realization?
3. What are my five best qualities?
4. What five areas of my life do I need to improve?
5. If Barbara Walters would ask me what my basic philosophy of life is, how would I answer?
6. Is my glass of water half full or half empty? Why?
7. What pervasive mood do I find myself in most of the time?
8. What do I think about people in general?
9. What role do my religion, culture, ethnic values, gender, and sexual orientation play in my view of life?
10. On the Meyers Briggs Type Inventory, what am I?
11. Who is my hero?
12. What is maturity?
13. What are my personal goals and objectives?
14. Who and/or what influenced my life?
15. What would my best friend say about me?
16. What is the biggest criticism people have of me?
17. What three adjectives best describe me?

WHO AM I AS A PROFESSIONAL?

1. What are my reasons for becoming a counselor?
2. Do I feel my emotional issues will be addressed and resolved by becoming a counselor?
3. What is my need to be a counselor?
4. What makes me think that I will be an effective counselor?
5. What are my countertransference issues?
6. What do I expect from clients?
7. What do I expect from my profession?
8. What do I anticipate getting from colleagues?
9. What are my professional strengths and weaknesses?

10. What are my professional goals and objectives?
11. What would my fellow students say about me?
12. With what type of clients do I wish to work and why?
13. How would I handle stress/burnout?
14. How do I handle praise and criticism of my work?

Every time you interact with your clients your professional and personal attributes and self become evident. As you build rapport with your clients, they come to value the way you present yourself. Clients want the assurance of your commitment to help them. The self-assessment process empowers you: the more you know about yourself, the more control you have over your life. This control can be passed along to clients. Refinements come about through continued interaction with colleagues, clients, and others with whom we have daily contact.

You will need to be careful not to expect too much of yourself too early in your career. For this reason we suggest being gentle with yourself; let the natural process of your professional development follow its course. Be yourself. Do not attempt to play a role, Jung's (1980) concept of a persona. Confidence should build over time. Your self-assessment enhances the formal skills, techniques, and theories you have learned in this evolving process.

Finally, knowing what you can and cannot do for your clients helps in the growth process and strengthens your knowledge base. The ACA Code of Ethics (B-12) encourages you to transfer clients to another professional if you determine an inability to be of assistance to your clients. You cannot be everything to everybody!

Being guide and mentor—and perceived rescuer—is not an easy task. The journey of life is not an easy one; nor is it fair. Our clients know this. At times it may be difficult for us as well as our clients to maintain a positive attitude toward humanity and even existence itself. Yet, we must not view life's journey as tedious, threatening, and dismal. We recall the words of a former colleague: "All we do in this lifetime is rearrange deck chairs on the Titanic." Is there more than this? We believe so. We have, however, no empirical proof, only a conviction that humanity is basically good and that we have some freedom to intervene in the condition of humankind, albeit limited. Maintaining a personal attitude of hope can only foster hope in our clients; exhibiting a basic trust in humankind can engender a sense of inward and profound trust in our clients; maintaining expectations for improvement can only serve as impetus for change; and demonstrating positive regard for those with whom we come in contact affirms and sustains them in their humanity and dignity. It is these beliefs for which we stand.

References

BENJAMIN, A. (1987). *The helping interview with case illustrations.* Boston: Houghton Mifflin.

BERNARD, J., & GOODYEAR, R. (1992). Clinical evaluation. In S. Badger (Ed.), *Fundamentals of clinical supervision* (pp. 105–130). Boston: Allyn & Bacon.

COREY, M., & COREY, J. (1993). *Becoming a helper* (2nd ed.). Pacific Grove, CA: Brooks/Cole.

DOWLING, S. (1984). Clinical evaluation: A comparison of self, self with videotape, peers, and supervisors. *The Clinical Supervisor, 2,* 71–78.

FREUD, S. (1943). *A general introduction to psychoanalysis.* Garden City, NY: Garden City Publishing.

JUNG, C. (1980). *The archetypes and the collective unconscious* (R. Hull, Trans.). Princeton, NJ: Princeton University Press. (Original work published 1902)

Bibliography

ATKINSON, D., WORTHINGTON, R., DANA, D., & GOOD, G. (1991). Etiology, beliefs, preferences for counseling orientations and counseling effectiveness. *Journal of Counseling Psychology, 38*(3), 258–264.

BALSAM, M., & BALSAM, A. (1984). *Becoming a psychotherapist.* Chicago: University of Chicago Press.

COREY, J. (1991). *Theory and practice of counseling and psychotherapy.* Pacific Grove, CA: Brooks/Cole.

PECK, M. (1978). *The road less traveled.* New York: Simon and Schuster.

SUSSMAN, M. (1992). *A curious calling: Unconscious motivations for practicing psychotherapy.* Northvale, NJ: Jason Aronson.

TRACEY, T. (1991). The structure of control and influence in counseling and psychotherapy: A comparison of several definitions and measures. *Journal of Counseling Psychology, 38*(3), 265–278.

VACHON, D., & AGRESTI, A. (1992). A training proposal to help mental health professionals clarify and manage implicit values in the counseling process. *Professional Psychology: Research and Practice, 23*(6), 509–514.

CHAPTER TEN

ETHICAL ISSUES

ETHICAL PRACTICE IN COUNSELING

The term *ethical* means that a behavior or practice conforms to the specialized rules and standards for proper conduct of a particular profession. Ethics, therefore, is "a fundamental and defining aspect of professionalism" (Gibson & Pope, 1993, p. 330). The code of ethics is basic to the identity of any professional organization, because it provides behavioral norms and guidelines that serve to structure each member's professional activities, as well as to ensure uniformly high professional standards among all members.

Counselors-in-training have an obligation to be cognizant of and to maintain an adherence to the Ethical Standards of the American Counseling Association (1988), in addition to any pertinent state codes and laws. These standards spell out distinct behavioral parameters within which all counselors, including yourself, as a counselor intern, must operate. The ACA Code of Ethics is divided into eight sections, each describing one aspect of proper professional behavior (see Appendix E for the complete ACA Code of Ethics).

Counselor interns also need to be aware of and to function within the boundaries of their agency or institutional policies, provided that these policies do not conflict with the ACA Code of Ethics. If you perceive a possible conflict, realize that the Code of Ethics takes precedence over agency or institutional policies. At that point, we suggest that you discuss the issue with your agency supervisor and university supervisor to try to resolve the problem. You may also wish to contact the ACA or the ethics committee of your state licensure or certification board or your state division of the ACA for further advice or assistance. Finally, as an intern and future professional counselor, you should always conduct your counseling activities in accordance with state and local laws and regulations. These rules can provide you with important professional backup and a sense of security in dealing with uncomfortable or equivocal professional situations.

ETHICAL DILEMMAS

Ethical codes provide guidelines and principles for ethical conduct, yet realistically, as counselors we will encounter many professional situations that are not specifically addressed by the ACA Code of Ethics (Corey, 1991; Corey, Corey, & Callanan, 1993; Gibson & Pope, 1993). Furthermore, certain clinical circumstances may involve conflicting interests or multiple ethical principles, which leave us with exceedingly difficult choices. "It is the translation of a code's principles into practical directions for conduct that is the greatest challenge for most of us" (Gibson & Pope, 1993, p. 330).

For example, one dilemma we have noticed in our own professional lives has been that in which personal morality comes into conflict with professional ethics. Our client may reveal to us that he or she is engaging in some behavior that we personally believe to be wrong or even immoral, yet we are bound by our ethical guidelines and the rules of professional confidentiality and cannot take any action outside of the treatment situation. The result can be a rather stubborn case of uncomfortable cognitive dissonance. One counselor intern agonized over a client who reported driving drunk several nights a week. Although he did encourage his client to work on this issue during their sessions, the intern told us, "If I could just be a private citizen and not a professional counselor, I would make a telephone call to report this person immediately! I feel a moral responsibility to keep her off the road."

Another area of potential professional quandary involves, if you will, choice of "client." For example, if you feel that a client is antisocial, presents a possibility of harm to others, or poses an outright danger to society, is that person your client or is the system or the general public your client? Where does your responsibility lie? One counselor intern described her dilemma when a client who was HIV positive disclosed that his partners were unaware of this test result and that he frequently did not adhere to safe sex practices. This intern wondered about her duty to warn and protect the partners versus her duty to protect client confidentiality. Another counselor intern described a situation in which she was counseling an undergraduate student with a history of violent, aggressive behavior, which was unknown to university administrators. This undergraduate was granted permission to live in a dormitory on campus. The counselor intern worried about the safety of the other students, yet felt that she could not warn university officials because the client had made no specific threats to hurt anyone.

GETTING HELP IN RESOLVING ETHICAL PROBLEMS

During your internship, and throughout your counseling career, you will grapple with difficult decisions as you try to apply the ACA Code of Ethics to ambiguous or complex counseling situations. We recommend using a variety of resources for help in resolving your concerns, including supervisors and colleagues, the ACA Ethics Committee, state licensure or certification boards, and professional journals and texts that address ethics in counseling. In their book *Issues and Ethics in the Helping Professions*, Corey, Corey, and Callanan (1993) analyze many of the ethical dilemmas counseling professionals may find themselves confronting. We encourage that you read this text thoroughly as a preventive measure against violation of standards of accepted practice and that you refer to it in any questionable professional circumstances, as well.

As a counseling professional, your own values and morals, as well as your personal conceptualization of ethical counseling behavior, will influence your interpretation and application of the ACA Code of Ethics. Welfel and Kitchener (1992) suggested that "the limitations of professional codes can be addressed by considering more fundamental ethical principles" (p. 180). The work of Welfel and Kitchener may help to provide a framework for you as you strive to make ethical decisions in counseling. They wrote that the five basic principles underlying ethical codes in the helping professions include:

1. Benefit others (implying a responsibility to do good).
2. Do no harm to others.
3. Respect the autonomy of others (including freedom of thought and freedom of action).
4. Act in a fair and just manner (meaning that the rights and interests of one individual must be balanced against the rights and interests of others).
5. Be faithful (implying trustworthiness, loyalty, and keeping promises). (Welfel & Kitchener, 1992)

Keeping these factors in mind may serve to clarify your understanding of those ethical dilemmas in counseling where the ACA Code of Ethics does not seem to provide definitive answers.

A BRIEF SUMMARY OF THE ACA CODE OF ETHICS

The first step in upholding the ethical standards and practices of the counseling profession during your internship is gaining a working knowledge of the ACA Code of Ethics. Many counseling students tend to avoid a thorough reading and/or a detailed examination of the ACA Code of Ethics. They feel that it is too complicated, too lengthy, and inaccessible, because it is often relegated to an appendix at the back of a book, rather than being included within the main text. In addition, the ACA Code of Ethics may seem somewhat remote from their immediate needs and frame of reference as graduate students, rather than as counseling practitioners. However, as a counselor intern, your role has evolved from that of primarily a student to that of essentially a clinician. We have therefore provided the following summary of the ACA Code of Ethics in an effort to make it more "user friendly." We strongly recommend that you set aside the time, early in your internship, to study the ACA Code of Ethics in its original, unabbreviated form, as well. A solid understanding of the ACA Code of Ethics will provide you with a knowledge base as you struggle to apply ethical principles to the diversity of challenging situations you will encounter during your counseling internship and your counseling career.

Section A: General

Counselors are responsible for upholding the ethical standards of the counseling profession by seeking to provide the highest quality of professional services in all settings. Counselors expect high standards from the institutions where they work, as well as from their professional associates, and they take necessary steps to ensure that ethical practices are in effect. All counseling services and products, including teaching and other types of communication media, are subject to these same ethical standards. Counselors recognize the need for continued professional development and do not provide services outside their area of training and competencies. They are careful and honest in representing their professional qualifications. Counselors recognize, respect, and honor the counselor–client relationship and protect the client's privacy, rights, and dignity at all times.

Section B: Counseling Relationship

Counselors always respect the rights, the freedom, the integrity, and the welfare of the client, whether in an individual relationship or in a group. Counselors explain the goals, structure, expectations, rules, treatment techniques, and any limitations of counseling to clients at the outset of the counseling relationship. Clients are always informed of any experimental treatments. Dual relationships are avoided (for example, if the prospective counselor/client are also teacher/student, supervisor/supervisee, or friends), and the client is referred for counseling elsewhere. Counselors do not enter into a professional relationship with a client who is already involved in another counseling relationship, unless the first professional is contacted and grants approval. If a counselor believes it advisable to terminate a counseling relationship, the counselor refers the client to a satisfactory alternative. Counselors take responsible action to protect individuals, including consulting with other professionals, if they feel that the client or others are in danger. Counselors view all sexual intimacies and relationships with clients as inappropriate and unethical.

Counselors assume responsibility for protecting group members from psychological or physical suffering resulting from participation in the group. Counselors protect group members by screening prospective members, by maintaining an awareness of group interactions and relationships, and by making available professional assistance for members who require such intervention, both during and after group sessions. Counselors advise group members regarding standards of confidentiality concerning other members. Counselors maintain confidentiality by taking precautions in their storage and disposal of client records. Counselors are careful to keep test data, tape recordings, and similar information used in counseling separate from the client's official records at the agency or institution. Counselors maintain confidentiality and do not share client information without the client's consent. Counselors limit computer storage of information as much as possible, destroy data no longer needed, and limit accessibility of computer data to others, by using computer security methods. Counselors always disguise identifying information when using material related to clients for research or teaching. When using computers as part of counseling services, counselors take precautions to make certain that the client is intellectually, physically, and emotionally able to use the computer, that the program is appropriate, that the client understands the purpose and possible limitations of the program, and that follow-up counseling is provided. Counselors who design computer software ensure that self-help/stand-alone software has actually been designed as such, rather than modified from preexisting software that required professional counselor assistance for use. Counselors designing such software also provide descriptions of expected outcomes, suggestions for use, inappropriate applications, possible benefits, and a manual defining other essential data.

Section C: Measurement and Evaluation

Counselors adhere to the highest standards in the selection, administration, and interpretation of all educational and psychological tests. Counselors understand that testing provides only one source of information for understanding and evaluating the client, and counselors are well aware of the limitations of tests, including the effect of socioeconomic, ethnic, and cultural factors. Counselors consider reliability and validity data, individual client needs, the currency or obsolescence of the test, and their own familiarity, competency, and experience with a particular test when choosing assessment instruments. When interpreting test results or when making decisions based on test results, counselors take care that they have adequate knowledge concerning psychological and educational assessment techniques, validity and reliability statistics, and test research. Counselors make certain that standardized tests are administered under the specified conditions that were designated and that test security is strictly respected (if applicable). Counselors explain the purpose of testing to the client and share test results with the client. Counselors adhere to guidelines for the preparation, publication, and distribution of tests as delineated in the three publications named in the ACA Code of Ethics (refer to Appendix E, Section C, Parts 16a, 16b, and 16c).

Section D: Research and Publication

Counselors who do research with human subjects adhere to the ethical principles and regulations described in five sources noted in the ACA Code of

Ethics (see Appendix E, Section D, Parts 1a through 1e). The principal researcher holds the primary responsibility for ethical practice in all research activities; however, all counselors involved in a project must be aware of ethical principles, must assume ethical responsibility for their own actions, and must take precautions to protect the physical, psychological, and social welfare of the human subjects.

Counselors inform all voluntary research subjects of the purpose of the investigation, if at all possible, and the identity of all subjects is carefully disguised. Counselors use involuntary subjects only if they are certain the research will not harm the subjects in any way whatsoever. Counselors report research findings in an ethical manner: conditions that might affect results are noted, and the research is carried out with the goal of minimizing misleading results; original research data is reported so that the research can be replicated; previous research in a similar area is cited, and the investigators or authors are given credit; research findings are reported even in the event that they reflect adversely on institutions, programs, services, or vested interests.

Counselors who publish written material are careful to give proper credit to the work of others through the use of citations, references, footnotes, acknowledgment, joint authorship, or other accepted means. Counselors who work together with other individuals on a research study or publication are responsible for the accuracy and the thoroughness of the information they contribute. In addition, counselors who collaborate incur an ethical responsibility to the other collaborators to be punctual and to perform as agreed. Counselors do not submit the same piece of written work to more than one journal or publisher at a time. Counselors obtain permission from the original publisher for reprinting manuscripts that have been previously published.

Section E: Consulting

Counselors understand that consultation involves finding solutions to issues, rather than counseling individual persons, with a focus on individual or organizational problems related to the workplace. Counselors providing such assistance collaborate with the client in determining the problem, the goals, the methods used to facilitate change, and the expected outcomes. Counselors strive to encourage adaptability and autonomy and to discourage dependence in their consultation clients. Counselors who consult are aware of their own values, knowledge, skills, and competencies, as well as their own limitations in providing services. Counselors therefore use appropriate referrals and resources for consultation at all times. Counselors always adhere to the ethical standards in remuneration and advertising practices.

Section F: Private Practice

Counselors engaged in the private practice of counseling adhere to ethical standards, as well as to state and local laws and regulations regarding such practice.

Counselors are careful in advertising their services and in providing accurate educational and professional information to clients; they avoid all misleading or deceptive material. Counselors do not allow their names to be used by a private professional organization unless they are actively engaged in counseling there. Counselors who are members of a partnership or corporation list professional competencies and specialties by name, in compliance with local regulations. Counselors do not use agency or institutional affiliations to attract clients for private counseling practice.

Section G: Personnel Administration

Counselors who work in public or semipublic institutions must share responsibility and accountability for setting and achieving goals, for formulating and implementing personnel policies, for ensuring and enhancing the rights and welfare of all clients, and for avoiding any practices that are inhumane, illegal, or unjustifiable with regard to hiring, promotion, or training. Counselors in such settings determine and specify the parameters of their professional roles, including their competencies, their interpersonal working agreements with other staff members, maintenance of confidentiality, work load, care of records, and accountability. Counselors in such settings alert employers to any conditions that may possibly be detrimental to ethical practices, submit to periodic professional evaluation, develop in-service programs, and assign other staff members only those tasks that are compatible with their competencies, training, and experience.

Counselors in public or semipublic settings disclose the limits of confidentiality and the use of supervision to clients at the outset of the counseling relationship.

Section H: Preparation Standards

Counselors who are responsible for training counseling students are competent practitioners and teachers who assume the highest ethical responsibilities of the profession. Counselors who train others strive to develop in their students an awareness of professional ethical behavior, ethical responsibilities, and ideals of humanitarian service. Counselor educators advise prospective students concerning program requirements, scope of professional skills development, and potential employment prior to admission to the counseling program. Counselor educators work toward the development of counseling programs that conform to the current educational guidelines of the ACA; integrate academic study with practical training; develop students' competencies, knowledge base, and self-awareness; include training in research and evaluation appropriate to the level of the students; expose students to diverse theoretical orientations so that students have the opportunity to make comparisons; and provide explicit policies with regard to all aspects of field placements. Counselor educators recognize and identify the levels of competencies of their students according to Division standards for paraprofessional as well as professional personnel. Counselors who train others evaluate their students and become aware of their limitations, so that these students can receive necessary remedial training or be removed from the program if necessary. Counselor educators whose programs include learning experiences that emphasize personal disclosure of intimate or sensitive material advise students of these experiences prior to program admission. In these cases, counselor educators involved in such experiential learning formats do not evaluate, supervise, or have administrative authority regarding the students participating in the experience. Students are always offered an alternative growth experience and are not penalized in any way for selecting these alternatives.

A MODEL FOR ETHICAL COUNSELING PRACTICE

Ethical behavior requires more than a familiarity with the profession's code of ethics. Often, implementing ethical decisions with ethical action involves substantial personal and professional risk or discomfort (Gawthrop & Uhleman, 1992). The motivation to accept these possible consequences to self of ethical actions may be considered a moral decision. Welfel and Kitchener (1992) presented a model for moral behavior, which consists of four components, each a psychological process. They considered this model from the perspective of the professional helper, and we suggest that their ideas have great merit for you as a counselor intern and for all professional counselors who are concerned with ethical practice. The components of moral and ethical behavior described by Welfel and Kitchener (1992) include:

1. *Interpreting the situation as a moral one* (recognizing the existence of an ethical problem and acknowledging that its outcome affects the welfare of your client and other people)
2. *Using moral reasoning* (understanding and applying ethical codes; using ethical principles where applicable)
3. *Deciding what one intends to do* (having recognized that an ethical problem exists, and having decided what the ethical response might be, the professional decides whether to proceed, despite the discomfort of confrontation or other ethical action)
4. *Implementing moral action* (persevering with ethical behavior despite costs to self or great pressure to withdraw; for example, being steadfast in the face of legal action or disapproval from colleagues)

CONCLUDING REMARKS

You will surely be confronted with a broad scope of ethical issues and dilemmas during your counseling internship and throughout your professional career. A knowledge of the ACA Code of Ethics provides you with a foundation for your ethical decisions; however, you will need to examine your own values closely as you search for acceptable solutions to some of these ethical problems. Corey (1991) explained:

> Although practitioners know the ethical standards of their professional organizations, they exercise their own judgment in applying these principles to particular cases. They realize that many problems are without clear-cut answers, and they accept the responsibility of searching for those answers (p. 82).

We encourage you to discuss ethical questions and issues with your colleagues, supervisors, and professors and to refer to the professional journals and books that address counseling ethics as you search for those answers.

References

COREY, G. (1991). *Theory and practice of counseling and psychotherapy.* Pacific Grove, CA: Brooks/Cole.

COREY, G., COREY, M., & CALLANAN, P. (1993). *Issues and ethics in the helping professions* (4th ed.). Pacific Grove, CA: Brooks/Cole.

ETHICAL STANDARDS OF THE AMERICAN COUNSELING ASSOCIATION (rev. ed.). (1988). Alexandria, VA: American Association of Counseling and Development Governing Council.

GAWTHROP, J., & UHLEMAN, M. (1992). Effects of the problem-solving approach in ethics training. *Professional Psychology: Research and Practice, 23*(1), 38–42.

GIBSON, W., & POPE, K. (1993). The ethics of counseling: A national survey of certified counselors. *Journal of Counseling and Development, 71*(3), 330–336.

WELFEL, E., & KITCHENER, K. (1992). Introduction to the special section: An agenda for the 90's. *Professional Psychology: Research and Practice, 23*(3), 179–181.

Bibliography

ARTHUR, G., & SWANSON, C. (1993). *Confidentiality and privileged communication.* Alexandria, VA: American Counseling Association.

BULLIS, R. (1992). *Law and management of a counseling agency or private practice.* Alexandria, VA: American Counseling Association.

BURKE, M., & MIRANTI, J. (1992). *Ethical and spiritual values in counseling.* Alexandria, VA: American Counseling Association.

COREY, G. (1991). *Manual for theory and practice of counseling and psychotherapy.* Pacific Grove, CA: Brooks/Cole.

GABBARD, G. (Ed.). (1989). *Sexual exploitation in professional relationships.* Washington, DC: American Psychiatric Press.

GOODYEAR, R., CREGO, C., & JOHNSTON, M. (1992). Ethical issues in the supervision of student research: A study of critical incidents. *Professional Psychology: Research and Practice, 23*(2), 203–210.

GUTHEIL, T., & GABBARD, G. (1993). The concept of boundaries in clinical practice: Theoretical and risk-management dimensions. *American Journal of Psychiatry, 150*(2), 188–196.

HARDING, A., GRAY, L., & NEAL, M. (1993). Confidentiality limits with clients who have HIV: A review of ethical and legal guidelines and professional policies. *Journal of Counseling and Development, 71*(3), 297–305.

HERLIHY, B., & GOLDEN, L. (1990). *Ethical standards casebook* (4th ed.). Alexandria, VA: American Counseling Association.

HOPKINS, B., & ANDERSON, B. (1990). *The counselor and the law* (3rd ed.). Alexandria, VA: American Counseling Association.

KITCHENER, K. (1992). Psychologist as teacher and mentor: Affirming ethical values throughout the curriculum. *Professional Psychology: Research and Practice, 70*(3), 190–195.

MacNAIR, R. (1992). Ethical dilemmas of child abuse reporting: Implications for mental health counselors. *Journal of Mental Health Counseling, 14*(2), 127–136.

MITCHELL, R. (1991). *Documentation in counseling records.* Alexandria, VA: American Counseling Association.

POPE, K., & VETLER, V. (1992). Ethical dilemmas encountered by members of the American Psychological Association: A national survey. *American Psychologist, 47*(3), 397–411.

ROSENBAUM, M. (Ed.). (1982). *Ethics and values in psychotherapy: A guidebook.* New York: Free Press.

RUTTER, P. (1989). *Sex in the forbidden zone.* New York: Fawcett.

SALO, M., & SHUMATE, S. (1993). *Counseling minor clients.* Alexandria, VA: American Counseling Association.

STEVENS-SMITH, P., & HUGHES, M. (1993). *Legal issues in marriage and family counseling.* Alexandria, VA: American Counseling Association.

VASQUEZ, M. (1992). Psychologist as clinical supervisor: Promoting ethical practice. *Professional Psychology: Research and Practice, 70*(2), 196–202.

WELFEL, E. (1992). Psychologist as ethics educator: Successes, failures, and unanswered questions. *Professional Psychology: Research and Practice, 70*(3), 182–189.

CHAPTER ELEVEN

FINISHING UP

We anticipate that it is with great pride and a sense of accomplishment that you now approach the end of your placement. We hope that your journey in clinical growth has been a fruitful one and that you are now ready to enter the "real world" of therapy.

You have, however, several items in need of attention before exiting this pre-professional experience. We have compiled a checklist of activities pertaining to clients' services, their continuity of care, and programmatic issues that will help you finish up the process.

Client Transfer

- Determine when you will notify your clients of the impending completion of your internship.
- Address clients' feelings and provide opportunities to say goodbyes.
- Arrange for a new counselor to be assigned.
- Schedule a counseling session for your clients to be introduced to their new counselor.
- Decide with your clients the stage of your counseling relationship conducive for the transfer to occur.
- Maintain precise documentation permitting the new counselor and client to know exactly where to begin services (treatment plan, transfer summary, progress notes).
- Notify the social service agencies that provided case management/advocacy of your impending departure and who will replace you.
- Clarify the current status of issues you have been addressing with various outside agencies.
- Verify that all letters, reports, and records that are requested and/or needed are sent.
- Have the client sign new releases of information so exchange of information continues.

Termination of Clients

- Attempt to achieve the goals of treatment plans for clients to be terminated.

- Notify clients of services available to them if the need arises (e.g., support groups, crisis telephone helplines, emergency counseling services).
- Adhere to the detailed provisions of after-care services (re: medication, review appointments, support groups, bibliotherapy, etc.).

Program Coverage

- Arrange for coverage for groups.
- Remove yourself from schedules of intake assessments, emergency services, didactic programs, and research activities.
- Curtail all client and program services within two weeks of departure from the site.
- Offer your telephone number and address if agency needs to contact you.

You should schedule an exit interview with your site supervisor and your university advisor to review your performance and to complete the final evaluation. You will discuss your areas of competencies (therapeutic orientation, theories, techniques), the clinical population with which you were most effective, and the service areas (individual, group, intake, emergency, case management) in which you demonstrated the most skill. General impressions and recommendations should be discussed at this time. Your input on your performance is to be considered as well as sufficient opportunities for you to ask questions and get any needed clarifications. Your supervisor should have been providing feedback on your performance during the entire placement; thus the final evaluation should not be a time for surprises. Your comments regarding your supervisor's role and his or her relationship with you can be discussed. Ideally, your comments on the internship experience will be welcomed and will only add to your supervisor's effectiveness.

Your site supervisor and university supervisor will, in all probability, endorse your successful completion of the internship in writing on an evaluation form. At the conclusion of this meeting, take the opportunity to complete any necessary licensure or certification forms.

Finally, your supervisor may recommend ways to present your experiences in a clear and concise manner in your résumé. As a practitioner, your supervisor has a greater understanding of what would increase your marketability. He or she may share information about employment opportunities and career trends. The supervisor may suggest that you "network" with mental health professionals regarding job opportunities and strategies. This would also be a good time to request a professional reference.

Recognize that you have an obligation to yourself to examine your own feelings regarding this termination. After all, has this internship not been a developmental process for you? Certainly, your clients have moved through a therapeutic process, but you yourself have also made an educational journey consisting of several discrete steps, including starting, acclimating (or adjusting), working, assessing, integrating, and finishing.

- *Starting:* You bring a fund of knowledge, enthusiasm, an eagerness to learn, an openness to experience our profession in the real world.

- *Adjusting*: You adapt to a new environment and individuals with the goal of understanding how you fit into the overall dynamics of the agency.
- *Working*: You are now able to use your knowledge and skills. You aspire to establish credibility as a counselor.
- *Assessing*: You evaluate the quality of your performance and test out your understanding of the intricacies of the counseling field. In addition, you solicit feedback regarding your efforts.
- *Integrating*: You are able to implement the feedback you receive, coupled with new practical knowledge gained, to become a more effective counselor.
- *Finishing*: You feel prepared to transfer the formal and practical knowledge into the counseling field.

We hope that this book has been helpful throughout your internship. We have tried to provide the practical information you needed to function in your role as counselor-in-training. We also trust that with this new source of knowledge you were able to integrate your theoretical training with the practical expectations and demands of our profession.

We like our clients to take a neatly wrapped package away with them from therapy, a sort of gift to themselves, one that they can feel good about. This package contains much hard work, personal growth, independence, a sense of accomplishment and survival, and, in Tillich's (1952) words, "the courage to be." Likewise, we want you to take a neatly wrapped package away with you as you complete this field experience. Listed below are the contents of this, your internship gift to yourself. Please permit us to wax philosophical.

- May you have the sense that you have participated in a meaningful professional, educational experience that has touched you personally.
- Emerson once wrote that one criterion for the determination of success is the knowledge that "by your having lived someone has breathed easier." May you come away with a commitment to assist others in their independence from you.
- May you develop a sense of competency: competency in skills, interactions, theoretical knowledge base, and asking for information and help when you do not know the answers.
- May you continue to grow and change and adapt to new and uncertain circumstances, in both your personal and your professional life, which are, after all, inextricably intertwined.
- May you leave a positive mark on others, as we hope we have done with you.

We wish you much success and happiness in our profession.

References

TILLICH, P. (1952). *The courage to be*. New Haven, CT: Yale University Press.

APPENDIX A

SAMPLE RÉSUMÉ

CHRIS JONES
2300 Main Street
Anytown, OH 44444
(216) 555-1111

Objective	An internship in the human services and counseling area.
Education	B.A. Hiram College, June, 1988 Major: Psychology Cumulative GPA: 3.5
	M.A. in Counseling, John Carroll University, Expected, May, 1997

Work experience	*Family Counseling Associates*, Cleveland, OH Case Aide	1990–1992
	Cleveland Memorial Hospital, Cleveland, OH Behavioral Technician	1988–1990
	McDonald's Restaurants, Chagrin Falls, OH Counter person	1984–1988

Honors and activities	Dean's List, Hiram College, 1985–1988 Tutoring Program, Cleveland, 1986 Psychology Club, Hiram College, 1987–1988 Presenter, Ohio Counseling Association, 1988
References	Available upon request

APPENDIX B

SAMPLE THANK-YOU NOTE

A. Smith, Ph.D.
Cleveland Counseling Agency
22000 Euclid Avenue
Cleveland, OH 44120

Dear Dr. Smith:

Thank you for taking time to meet with me yesterday. I enjoyed having the opportunity to speak with you and to learn about your agency.

I am interested in a field placement with the Cleveland Counseling Agency beginning in September 1993. I will call you in about a week to discuss this possibility.

Sincerely,

Chris Jones

APPENDIX C

SAMPLE INTERNSHIP ANNOUNCEMENT

INTERNSHIP/PRACTICUM PLACEMENT
FOR GRADUATE STUDENTS IN COUNSELING
Available for the Fall Semester
Cleveland County Mental Health Center

Opportunities for:

- Mental Health Counseling with a wide range of clients
- Participation in current aspects of the Agency's Outpatient and Community Education programs
- Development of new services for clients

Training supervision toward Ohio licensure in counseling available.

Contact: A. Smith, Ph.D.
Clinical Director
(216) 555-2200

APPENDIX D

NBCC COURSEWORK REQUIREMENTS

NBCC COURSEWORK AREA DESCRIPTIONS

1. *Counseling theory* includes studies of basic theories, principles, and techniques of counseling and their application to professional counseling settings.

2. *Practicum/internship* refers to supervised counseling experience in an appropriate work setting of at least one semester duration for academic credit. The practicum must be taken through a regionally accredited institution. Five years of post-master's counseling experience with supervision may be substituted for a 3-semester hour/5-quarter hour practicum.

3. *Human growth and development* includes studies that provide a broad understanding of the nature and needs of individuals at all developmental levels; normal and abnormal human behavior; personality theory; life-span theory; and learning theory within cultural contexts.

4. *Social and cultural foundations* includes studies that provide a broad understanding of societal changes and trends; human roles; societal subgroups; social mores and interaction patterns; multicultural and pluralistic trends; differing lifestyles; and major societal concerns including stress, person abuse, substance abuse, discrimination, and methods for alleviating these concerns.

5. *The helping relationship* includes studies that provide a broad understanding of philosophic bases of helping processes; counseling theories and their applications; basic and advanced helping skills; consultation theories and their applications; client and helper self-understanding and self-development; and facilitation of client or consultee change.

6. *Group dynamics, processing and counseling* includes studies that provide a broad understanding of group development, dynamics, and counseling theories; group leadership styles; basic and advanced group counseling methods and skills; and other group approaches.

7. *Lifestyle and career development* includes studies that provide a broad understanding of career development theories; occupational and educational information sources and systems; career and leisure counseling, guidance, and education; lifestyle and career decision making; and career development program planning, resources, and effectiveness evaluation.

8. *Appraisal of individuals* includes studies that provide a broad understanding of group and individual educational and psychometric theories and approaches to appraisal; data and information-gathering methods; validity and reliability; psychometric statistics; factors influencing appraisals; and use of appraisal results in helping processes. Additionally, the specific ability to administer and interpret tests and inventories to assess abilities, determine interests, and identify career options is important.

9. *Research and evaluation* includes studies that provide a broad understanding of types of research; basic statistics; research-report development; research implementation; program evaluation; needs assessment; publication of research information; and ethical and legal considerations.

10. *Professional orientation* includes studies that provide a broad understanding of professional roles and functions; professional goals and objectives; professional organizations and associations; professional history and trends; ethical and legal standards; professional preparation standards; and professional credentialing.*

*National Board for Certified Counselors (1994). General Practice Counselor Certification Coursework Descriptions, p. 24.

APPENDIX E

ACA CODE OF ETHICS

Ethical Standards of the American Counseling Association

(3rd revision, AACD Governing Council, March 1988)

PREAMBLE

The Association is an educational, scientific, and professional organization whose members are dedicated to the enhancement of the worth, dignity, potential, and uniqueness of each individual and thus to the service of society.

The Association recognizes that the role definitions and work settings of its members include a wide variety of academic disciplines, levels of academic preparation, and agency services. This diversity reflects the breadth of the Association's interest and influence. It also poses challenging complexities in efforts to set standards for the performance of members, desired requisite preparation or practice, and supporting social, legal, and ethical controls.

The specification of ethical standards enables the Association to clarify to present and future members and to those served by members the nature of ethical responsibilities held in common by its members.

The existence of such standards serves to stimulate greater concern by members for their own professional functioning and for the conduct of fellow professionals such as counselors, guidance and student personnel workers, and others in the helping professions. As the ethical code of the Association, this document establishes principles that define the ethical behavior of Association members. Additional ethical guidelines developed by the Association's Divisions for their specialty areas may further define a member's ethical behavior.

SECTION A: GENERAL

1. The member influences the development of the profession by continuous efforts to improve professional practices, teaching, services, and research. Professional growth is continuous throughout the member's career and is exemplified by the development of a philosophy that explains why and how a member functions in the helping relationship. Members must gather data on their effectiveness and be guided by the findings. Members recognize the need for continuing education to ensure competent service.

2. The member has a responsibility both to the individual who is served and to the institution within which the service is performed to maintain high standards of professional conduct. The member strives to maintain the highest

levels of professional services offered to the individuals to be served. The member also strives to assist the agency, organization, or institution in providing the highest caliber of professional services. The acceptance of employment in an institution implies that the member is in agreement with the general policies and principles of the institution. Therefore the professional activities of the member are also in accord with the objectives of the institution. If, despite concerted efforts, the member cannot reach agreement with the employer as to acceptable standards of conduct that allow for changes in institutional policy conducive to the positive growth and development of clients, then terminating the affiliation should be seriously considered.

3. Ethical behavior among professional associates, both members and nonmembers, must be expected at all times. When information is possessed that raises doubt as to the ethical behavior of professional colleagues, whether Association members or not, the member must take action to attempt to rectify such a condition. Such action shall use the institution's channels first and then use procedures established by the Association.

4. The member neither claims nor implies professional qualifications exceeding those possessed and is responsible for correcting any misrepresentations of these qualifications by others.

5. In establishing fees for professional counseling services, members must consider the financial status of clients and locality. In the event that the established fee structure is inappropriate for a client, assistance must be provided in finding comparable services of acceptable cost.

6. When members provide information to the public or to subordinates, peers, or supervisors, they have a responsibility to ensure that the content is general, unidentified client information that is accurate, unbiased, and consists of objective, factual data.

7. Members recognize their boundaries of competence and provide only those services and use only those techniques for which they are qualified by training or experience. Members should accept only those positions for which they are professionally qualified.

8. In the counseling relationship, the counselor is aware of the intimacy of the relationship and maintains respect for the client and avoids engaging in activities that seek to meet the counselor's personal needs at the expense of that client.

9. Members do not condone or engage in sexual harassment, which is defined as deliberate or repeated comments, gestures, or physical contacts of a sexual nature.

10. The member avoids bringing personal issues into the counseling relationship, especially if the potential for harm is present. Through awareness of the negative impact of both racial and sexual stereotyping and discrimination, the counselor guards the individual rights and personal dignity of the client in the counseling relationship.

11. Products or services provided by the member by means of classroom instruction, public lectures, demonstrations, written articles, radio or television programs, or other types of media must meet the criteria cited in these standards.

SECTION B: COUNSELING RELATIONSHIP

This section refers to practices and procedures of individual and/or group counseling relationships. The member must recognize the need for client freedom of choice. Under those circumstances where this is not possible, the member must apprise clients of restrictions that may limit their freedom of choice.

1. The member's primary obligation is to respect the integrity and promote the welfare of the client(s), whether the client(s) is (are) assisted individually or in a group relationship. In a group setting, the member is also responsible for taking reasonable precautions to protect individuals from physical and/or psychological trauma resulting from interaction within the group.

2. Members make provisions for maintaining confidentiality in the storage and disposal of records and follow an established record retention and disposition policy. The counseling relationship and information resulting therefrom must be kept confidential, consistent with the obligations of the member as a professional person. In a group counseling setting, the counselor must set a norm of confidentiality regarding all group participants' disclosures.

3. If an individual is already in a counseling relationship with another professional person, the member does not enter into a counseling relationship without first contacting and receiving the approval of that other professional. If the member discovers that the client is in another counseling relationship after the counseling relationship begins, the member must gain the consent of the other professional or terminate the relationship, unless the client elects to terminate the other relationship.

4. When the client's condition indicates that there is clear and imminent danger to the client or others, the member must take reasonable personal action or inform responsible authorities. Consultation with other professionals must be used where possible. The assumption of responsibility for the client's(s') behavior must be taken only after careful deliberation. The client must be involved in the resumption of responsibility as quickly as possible.

5. Records of the counseling relationship, including interview notes, test data, correspondence, tape recordings, electronic data storage, and other documents, are to be considered professional information for use in counseling, and they should not be considered a part of the records of the institution or agency in which the counselor is employed unless specified by state statute or regulation. Revelation to others of counseling material must occur only upon the expressed consent of the client.

6. In review of the extensive data storage and processing capacities of the computer, the member must ensure that data maintained on a computer is: (a) limited to information that is appropriate and necessary for the services being provided; (b) destroyed after it is determined that the information is no longer of any value in providing services; and (c) restricted in terms of access to appropriate staff members involved in the provision of services by using the best computer security methods available.

7. Use of data derived from a counseling relationship for purposes of counselor training or research shall be confined to content that can be disguised to ensure full protection of the identity of the subject client.

8. The member must inform the client of the purposes, goals, techniques, rules of procedure, and limitations that may affect the relationship at or before the time that the counseling relationship is entered. When working with minors or persons who are unable to give consent, the member protects these clients' best interests.

9. In view of common misconceptions related to the perceived inherent validity of computer-generated data and narrative reports, the member must ensure that the client is provided with information as part of the counseling relationship that adequately explains the limitations of computer technology.

10. The member must screen prospective group participants, especially when emphasis is on self-understanding and growth through self-disclosure. The member must maintain an awareness of the group participants' compatibility throughout the life of the group.

11. The member may choose to consult with any other professionally competent person about a client. In choosing a consultant, the member must avoid placing the consultant in a conflict of interest situation that would preclude the consultant's being a proper party to the member's efforts to help the client.

12. If the member determines an inability to be of professional assistance to the client, the member must either avoid initiating the counseling relationship or immediately terminate that relationship. In either event, the member must suggest appropriate alternatives. (The member must be knowledgeable about referral resources so that a satisfactory referral can be initiated.) In the event the client declines the suggested referral, the member is not obligated to continue the relationship.

13. When the member has other relationships, particularly of an administrative, supervisory, and/or evaluative nature, with an individual seeking counseling services, the member must not serve as the counselor but should refer the individual to another professional. Only in instances where such an alternative is unavailable and where the individual's situation warrants counseling intervention should the member enter into and/or maintain a counseling relationship. Dual relationships with clients that might impair the member's objectivity and professional judgment (e.g., as with close friends or relatives) must be avoided and/or the counseling relationship terminated through referral to another competent professional.

14. The member will avoid any type of sexual intimacies with clients. Sexual relationships with clients are unethical.

15. All experimental methods of treatment must be clearly indicated to prospective recipients, and safety precautions are to be adhered to by the member.

16. When computer applications are used as a component of counseling services, the member must ensure that: (a) the client is intellectually, emotionally, and physically capable of using the computer application; (b) the computer application is appropriate for the needs of the client; (c) the client understands the purpose and operation of the computer application; and (d) a follow-up of client use of a computer application is provided to both correct possible problems (misconceptions or inappropriate use) and assess subsequent needs.

17. When the member is engaged in short-term group treatment/training programs (e.g., marathons and other encounter-type or growth groups), the member ensures that there is professional assistance available during and following the group experience.

18. Should the member be engaged in a work setting that calls for any variation from the above statements, the member is obligated to consult with other professionals whenever possible to consider justifiable alternatives.

19. The member must ensure that members of various ethnic, racial, religious, disability, and socioeconomic groups have equal access to computer applications used to support counseling services and that the content of available computer applications does not discriminate against the groups described above.

20. When computer applications are developed by the member for use by the general public as self-help/stand-alone computer software, the member must ensure that: (a) self-help computer applications are designed from the beginning to function in a stand-alone manner, as opposed to modifying software that was originally designed to require support from a counselor; (b) self-help computer applications will include within the program statements regarding intended user outcomes, suggestions for using the software, a description of the conditions under which self-help computer applications might not be appropriate, and a description of when and how counseling services might be beneficial; and (c) the manual for such applications will include the qualifications of the developer, the development process, validation data, and operating procedures.

SECTION C: MEASUREMENT AND EVALUATION

The primary purpose of educational and psychological testing is to provide descriptive measures that are objective and interpretable in either comparative or absolute terms. The member must recognize the need to interpret the statements that follow as applying to the whole range of appraisal techniques including test and nontest data. Test results constitute only one of a variety of pertinent sources of information for personnel, guidance, and counseling decisions.

1. The member must provide specific orientation or information to the examinee(s) prior to and following the test administration so that the results of testing may be placed in proper perspective with other relevant factors. In so doing, the member must recognize the effects of socioeconomic, ethnic, and cultural factors on test scores. It is the member's professional responsibility to use additional unvalidated information carefully in modifying interpretation of the test results.

2. In selecting tests for use in a given situation or with a particular client, the member must consider carefully the specific validity, reliability, and appropriateness of the test(s). General validity, reliability, and related issues may be questioned legally as well as ethically when tests are used for vocational and educational selection, placement, or counseling.

3. When making any statements to the public about tests and testing, the member must give accurate information and avoid false claims or misconceptions.

Special efforts are often required to avoid unwarranted connotations of such terms as IQ and grade equivalent scores.

4. Different tests demand different levels of competence for administration, scoring, and interpretation. Members must recognize the limits of their competence and perform only those functions for which they are prepared. In particular, members using computer-based test interpretations must be trained in the construct being measured and the specific instrument being used prior to using this type of computer application.

5. In situations where a computer is used for test administration and scoring, the member is responsible for ensuring that administration and scoring programs function properly to provide clients with accurate test results.

6. Tests must be administered under the same conditions that were established in their standardization. When tests are not administered under standard conditions or when unusual behavior or irregularities occur during the testing session, those conditions must be noted and the results designated as invalid or of questionable validity. Unsupervised or inadequately supervised test-taking, such as the use of tests through the mails, is considered unethical. On the other hand, the use of instruments that are so designed or standardized to be self-administered and self-scored, such as interest inventories, is to be encouraged.

7. The meaningfulness of test results used in personnel, guidance, and counseling functions generally depends on the examinee's unfamiliarity with the specific items on the test. Any prior coaching or dissemination of the test materials can invalidate test results. Therefore, test security is one of the professional obligations of the member. Conditions that produce most favorable test results must be made known to the examinee.

8. The purpose of testing and the explicit use of results must be made known to the examinee prior to testing. The counselor must ensure that instrument limitations are not exceeded and that periodic review and/or retesting is made to prevent client stereotyping.

9. The examinee's welfare and explicit prior understanding must be the criteria for determining the recipients of the test results. The member must see that specific interpretation accompanies any release of individual or group test data. The interpretation of test data must be related to the examinee's particular concerns.

10. Members responsible for making decisions based on test results have an understanding of educational and psychological measurement, validation criteria, and test research.

11. The member must be cautious when interpreting the results of research instruments possessing insufficient technical data. The specific purposes for the use of such instruments must be stated explicitly to examinees.

12. The member must proceed with caution when attempting to evaluate and interpret the performance of minority group members or other persons who are not represented in the norm group on which the instrument was standardized.

13. When computer-based test interpretations are developed by the member to support the assessment process, the member must ensure that the validity of such interpretations is established prior to the commercial distribution of such a computer application.

14. The member recognizes that test results may become obsolete. The member will avoid and prevent the misuse of obsolete test results.

15. The member must guard against the appropriation, reproduction, or modification of published tests or parts thereof without acknowledgment and permission from the previous publisher.

16. Regarding the preparation, publication, and distribution of tests, reference should be made to:

    a. "Standards for Educational and Psychological Testing," revised edition, 1985, published by the American Psychological Association on behalf of itself, the American Educational Research Association, and the National Council on Measurement in Education.

    b. "The Responsible Use of Tests: A Position Paper of AMEG, APGA, and NCME," *Measurement and Evaluation in Guidance,* 1972, 5, 385–388.

    c. "Responsibilities of Users of Standardized Tests," APGA, *Guidepost,* October 5, 1978, pp. 5–8.

SECTION D: RESEARCH AND PUBLICATION

1. Guidelines on research with human subjects shall be adhered to, such as:

    a. Ethical Principles in the Conduct of Research with Human Participants, Washington, D.C., American Psychological Association, Inc., 1982.

    b. Code of Federal Regulation, Title 45, Subtitle A, Part 46, as currently issued.

    c. Ethical Principles of Psychologists, American Psychological Association, Principle #9: Research with Human Participants.

    d. Family Educational Rights and Privacy Act (the Buckley Amendment).

    e. Current federal regulations and various states' rights privacy acts.

2. In planning any research activity dealing with human subjects, the member must be aware of and responsive to all pertinent ethical principles and ensure that the research problem, design, and execution are in full compliance with them.

3. Responsibility for ethical research practice lies with the principal researcher, while others involved on the research activities share ethical obligation and full responsibility for their own actions.

4. In research with human subjects, researchers are responsible for the subjects' welfare throughout the experiment, and they must take all reasonable precautions to avoid causing injurious psychological, physical, or social effects on their subjects.

5. All research subjects must be informed of the purpose of the study except when withholding information or providing misinformation to them is

essential to the investigation. In such research the member must be responsible for corrective action as soon as possible following completion of the research.

6. Participation in research must be voluntary. Involuntary participation is appropriate only when it can be demonstrated that participation will have no harmful effects on subjects and is essential to the investigation.

7. When reporting research results, explicit mention must be made of all variables and conditions known to the investigator that might affect the outcome of the investigation or the interpretation of the data.

8. The member must be responsible for conducting and reporting investigations in a manner that minimizes the possibility that results will be misleading.

9. The member has an obligation to make available sufficient original research data to qualified others who may wish to replicate the study.

10. When supplying data, aiding in the research of another person, reporting research results, or making original data available, due care must be taken to disguise the identity of the subjects in the absence of specific authorization from such subjects to do otherwise.

11. When conducting and reporting research, the member must be familiar with and give recognition to previous work on the topic, as well as to observe all copyright laws and follow the principles of giving full credit to all to whom credit is due.

12. The member must give due credit through joint authorship, acknowledgment, footnote statements, or other appropriate means to those who have contributed significantly to the research and/or publication, in accordance with such contributions.

13. The member must communicate to other members the results of any research judged to be of professional or scientific value. Results reflecting unfavorably on institutions, programs, services, or vested interests must not be withheld for such reasons.

14. If members agree to cooperate with another individual in research and/or publication, they incur an obligation to cooperate as promised in terms of punctuality of performance and with full regard to the completeness and accuracy of the information required.

15. Ethical practice requires that authors not submit the same manuscript or one essentially similar in content for simultaneous publication consideration by two or more journals. In addition, manuscripts published in whole or in substantial part in another journal or published work should not be submitted for publication without acknowledgment and permission from the previous publication.

SECTION E: CONSULTING

Consultation refers to a voluntary relationship between a professional helper and help-needing individual, group, or social unit in which the consultant is providing help to the client(s) in defining and solving a work-related problem with a client or client system.

1. The member acting as consultant must have a high degree of self-awareness of his/her own values, knowledge, skills, limitations, and needs in entering a helping relationship that involves human and/or organizational change and that the focus of the relationship be on the issues to be resolved and not on the person(s) presenting the problem.

2. There must be understanding and agreement between member and client for the problem definition, change of goals, and prediction of consequences of interventions selected.

3. The member must be reasonably certain that she/he or the organization represented has the necessary competencies and resources for giving the kind of help that is needed now or may be needed later and that appropriate referral resources are available to the consultant.

4. The consulting relationship must be one in which client adaptability and growth toward self-direction are encouraged and cultivated. The member must maintain this role consistently and not become a decision maker for the client or create a future dependency on the consultant.

5. When announcing consultant availability for services, the member conscientiously adheres to the Association's Ethical Standards.

6. The member must refuse a private fee or other remuneration for consultation with persons who are entitled to these services through the member's employing institution or agency. The policies of a particular agency may make explicit provisions for private practice with agency clients by members of its staff. In such instances, the clients must be apprised of other options open to them should they seek private counseling services.

SECTION F: PRIVATE PRACTICE

1. The member should assist the profession by facilitating the availability of counseling services in private as well as public settings.

2. In advertising services as a private practitioner, the member must advertise the services in a manner that accurately informs the public of professional services, expertise, and techniques of counseling available. A member who assumes an executive leadership role in the organization shall not permit his/her name to be used in professional notices during periods when he/she is not actively engaged in the private practice of counseling.

3. The member may list the following: highest relevant degree, type and level of certification and/or license, address, telephone number, office hours, type and/or description of services, and other relevant information. Such information must not contain false, inaccurate, misleading, partial, out-of-context, or deceptive material or statements.

4. Members do not present their affiliation with any organization in such a way that would imply inaccurate sponsorship or certification by that organization.

5. Members may join in partnership/corporation with other members and/or other professionals provided that each member of the partnership or corporation makes clear the separate specialties by name in compliance with the regulations of the locality.

6. A member has an obligation to withdraw from a counseling relationship if it is believed that employment will result in violation of the Ethical Standards. If the mental or physical condition of the member renders it difficult to carry out an effective professional relationship or if the member is discharged by the client because the counseling relationship is no longer productive for the client, then the member is obligated to terminate the counseling relationship.

7. A member must adhere to the regulations for private practice of the locality where the services are offered.

8. It is unethical to use one's institutional affiliation to recruit clients for one's private practice.

SECTION G: PERSONNEL ADMINISTRATION

It is recognized that most members are employed in public or quasi-public institutions. The functioning of a member within an institution must contribute to the goals of the institution and vice versa if either is to accomplish their respective goals or objectives. It is therefore essential that the member and the institution function in ways to: (a) make the institution's goals explicit and public; (b) make the member's contribution to institutional goals specific; and (c) foster mutual accountability for goal achievement.

To accomplish these objectives, it is recognized that the member and the employer must share responsibilities in the formulation and implementation of personnel policies.

1. Members must define and describe the parameters and levels of their professional competency.

2. Members must establish interpersonal relations and working agreements with supervisors and subordinates regarding counseling or clinical relationships, confidentiality, distinction between public and private material, maintenance and dissemination of recorded information, work load, and accountability. Working agreements in each instance must be specified and made known to those concerned.

3. Members must alert their employers to conditions that may be potentially disruptive or damaging.

4. Members must inform employers of conditions that may limit their effectiveness.

5. Members must submit regularly to professional review and evaluation.

6. Members must be responsible for in-service development of self and/or staff.

7. Members must inform their staff of goals and programs.

8. Members must provide personnel practices that guarantee and enhance the rights and welfare of each recipient of their service.

9. Members must select competent persons and assign responsibilities compatible with their skills and experiences.

10. The member, at the onset of a counseling relationship, will inform the client of the member's intended use of supervisors regarding the disclosure of information concerning this case. The member will clearly inform the client of the limits of confidentiality in the relationship.

11. Members, as either employers or employees, do not engage in or condone practices that are inhumane, illegal, or unjustifiable (such as considerations based on sex, handicap, age, race) in hiring, promotion, or training.

SECTION H: PREPARATION STANDARDS

Members who are responsible for training others must be guided by the preparation standards of the Association and relevant Division(s). The member who functions in the capacity of trainer assumes unique ethical responsibilities that frequently go beyond that of the member who does not function in a training capacity. These ethical responsibilities are outlined as follows:

1. Members must orient students to program expectations, basic skills development, and employment prospects prior to admission to the program.

2. Members in charge of learning experiences must establish programs that integrate academic study and supervised practice.

3. Members must establish a program directed toward developing students' skills, knowledge, and self-understanding, stated whenever possible in competency or performance terms.

4. Members must identify the levels of competencies of their students in compliance with relevant Division standards. These competencies must accommodate the paraprofessional as well as the professional.

5. Members, through continual student evaluation and appraisal, must be aware of the personal limitations of the learner that might impede future performance. The instructor must not only assist the learner in securing remedial assistance but also screen from the program those individuals who are unable to provide competent services.

6. Members must provide a program that includes training in research commensurate with levels of role functioning. Paraprofessional and technician-level personnel must be trained as consumers of research. In addition, personnel must learn how to evaluate their own and their program's effectiveness. Graduate training, especially at the doctoral level, would include preparation for original research by the member.

7. Members must make students aware of the ethical responsibilities and standards of the profession.

8. Preparatory programs must encourage students to value the ideals of service to individuals and to society. In this regard, direct financial remuneration or lack thereof must not influence the quality of service rendered. Monetary considerations must not be allowed to overshadow professional and humanitarian needs.

9. Members responsible for educational programs must be skilled as teachers and practitioners.

10. Members must present thoroughly varied theoretical positions so that students may make comparisons and have the opportunity to select a position.

11. Members must develop clear policies within their educational institutions regarding field placement and the roles of the student and the instructor in such placement.

12. Members must ensure that forms of learning focusing on self-understanding or growth are voluntary, or if required as part of the educational program, are made known to prospective students prior to entering the program. When the educational program offers a growth experience with an emphasis on self-disclosure or other relatively intimate or personal involvement, the member must have no administrative, supervisory, or evaluating authority regarding the participant.

13. The member will at all times provide students with clear and equally acceptable alternatives for self-understanding or growth experiences. The member will assure students that they have a right to accept these alternatives without prejudice or penalty.

14. Members must conduct an educational program in keeping with the current relevant guidelines of the Association.*

*ACA Code of Ethics (1988). Reprinted by permission of the American Counseling Association.

APPENDIX F

STATE AND NATIONAL CREDENTIALING BOARDS

STATE CREDENTIALING BOARDS

ALABAMA
Board of Examiners in Counseling
P.O. Box 550397
Birmingham, AL 35255
(205) 933-8100

ARIZONA
Counselor Credentialing Committee
of the Board of Behavioral Examiners
1645 W. Jefferson, 4th Floor
Phoenix, AZ 85007
(602) 542-1882

ARKANSAS
Board of Examiners in Counseling
Southern Arkansas University
P.O. Box 1396
Magnolia, AR 71753
(501) 235-4052

CALIFORNIA
Board of Behavioral Science Examiners
400 R Street, Suite 3150
Sacramento, CA 95814-6240
(916) 445-4933

COLORADO
Board of Licensed Professional
Counselor Examiners
1560 Broadway, Suite 1340
Denver, CO 80202
(303) 894-7766

DELAWARE
Board of Professional Counselors of
Mental Health
P.O. Box 1401
Margaret O'Neill Bldg.
Dover, DE 19903
(302) 739-4522

DISTRICT OF COLUMBIA
DC Board of Professional Counselors
Room LL-202, 605 G Street, NW
Washington, DC 20001
(202) 727-7454

FLORIDA
Board of Clinical Social Workers,
Marriage & Family Therapists, &
Mental Health Counselors

Florida Dept. of Professional Regulation
1940 N. Monroe Street
Tallahassee, FL 32399-0753
(904) 487-2520

GEORGIA
Composite Board of Professional
Counselors, Social Workers, and
Marriage & Family Therapists
166 Pryor Street, SW
Atlanta, GA 30303
(404) 656-3933

IDAHO
Idaho State Counselor Licensure Board
Bureau of Occupational Licenses
1109 Main Street, Suite 220
Boise, ID 83702-5642
(208) 334-3233

ILLINOIS
Counseling Board
6933 Blue Flag
Woodbridge, IL 60517
(708) 241-1191

IOWA
Board of Behavioral Science Examiners
2507 University Memorial, 307
Des Moines, IA 50311
(515) 242-5937

KANSAS
Behavioral Sciences Regulatory Board
Landon Office Bldg., Room 651-S
900 SW Jackson
Topeka, KS 66612-1263
(913) 296-3240

LOUISIANA
Licensed Professional Counselors
Board of Examiners
4664 Jamestown Avenue, Suite 125
Baton Rouge, LA 70808-3218
(504) 922-1499

MAINE
Board of Counseling Professionals
State House
Station N-35

Augusta, ME 04333
(207) 582-8723

MARYLAND
Board of Examiners of Professional
Counselors
Metro Executive Center, 3rd Floor
4201 Patterson Avenue
Baltimore, MD 21215-22299
(410) 764-4732

MASSACHUSETTS
Board of Allied Mental Health &
Human Service Professionals
100 Cambridge Street, 15th Floor
Boston, MA 02202
(617) 727-1716

MICHIGAN
Board of Professional Counselors
P.O. Box 30018
Lansing, MI 48909
(517) 335-0918

MISSISSIPPI
Board of Examiners for LPCs
P.O. Drawer 6239
Mississippi State, MS 39762-6239
(601) 325-8182

MISSOURI
Missouri Committee for Professional
Counselors
P.O. Box 162
Jefferson City, MO 65102
(314) 751-0018

MONTANA
Board of Social Work Examiners &
Professional Counselors
Arcade Bldg., Lower Level
111 North Jackson
P.O. Box 200513
Helena, MT 59620-0513
(406) 444-4285

NEBRASKA
Board of Examiners in
Professional Counseling
Bureau of Examining Boards

301 Centennial Mall South
P.O. Box 95007
Lincoln, NE 68509-5007
(402) 471-2115

NEW HAMPSHIRE
NH Board of Examiners of Psych.
& Mental Health Practice
105 Pleasant Street #457
Concord, NH 03301
(603) 226-2599

NEW MEXICO
P.O. Box 25101
Santa Fe, NM 87504
(505) 827-7197

NORTH CAROLINA
Board of Registered Practicing
Counselors
P.O. Box 12023
Raleigh, NC 27605
(919) 870-9557

NORTH DAKOTA
Board of Counselor Examiners
P.O. Box 2735
Bismarck, ND 58502
(701) 224-8234

OHIO
Counselor & Social Worker Board
77 South High Street, 16th Floor
Columbus, OH 43266-0340
(614) 466-0912

OKLAHOMA
Licensed Professional Counselors
Licensed Marital & Family Therapists
1000 NE 10th Street
Oklahoma City, OK 73117-1299
(405) 271-6030

OREGON
Board of Licensed Professional
Counselors and Therapists
796 Winter Street, NE
Salem, OR 97310
(503) 378-5499

RHODE ISLAND
Boards of Mental Health Counselors
& M&F Therapists
Division of Professional Regulation
3 Capitol Hill
Cannon Bldg., Room 104
Providence, RI 02908-5097
(401) 277-2827

SOUTH CAROLINA
Board of Examiners in Counseling
P.O. Box 7965
Columbia, SC 29202
(803) 734-1765

SOUTH DAKOTA
South Dakota Board of Counselor
Examiners
P.O. Box 1115
Pierre, SD 57501
(605) 224-6281

TENNESSEE
State Board of Professional
Counselors & M&F Therapists
283 Plus Park Blvd.
Nashville, TN 37247-1010
(615) 367-6280

TEXAS
Board of Examiners of Professional
Counselors
1100 W. 49th Street
Austin, TX 78756-3183
(512) 834-6658

VERMONT
CCMHC Advisory Board
Office of Professional Regulation
109 State Street
Montpelier, VT 05609-1106
(802) 828-2390

VIRGINIA
Board of Professional Counselors
Dept. of Health Professions
6606 W. Broad Street, 4th Floor
Richmond, VA 23230-1717
(804) 662-9912

WASHINGTON
Dept. of Health
Professional Licensing Services
Division
P.O. Box 47869
Olympia, WA 98504-7869
(206) 753-6936

WEST VIRGINIA
Board of Examiners in Counseling
P.O. Box 6492
Charlestown, WV 25362
(304) 345-3852

WISCONSIN
Board of Examiners of Certified
Professional Counselors

County Highway P W 8292
Pardeeville, WI 53954
(608) 742-3636

WYOMING
Professional Counselors, M&F
Therapists, Social Workers &
Chemical Dependency Specialists
Licensing Board
2301 Central Avenue
Barrett Bldg., 3rd Floor
Cheyenne, WY 82002
(307) 777-7788

NATIONAL COUNSELOR CERTIFICATION

National Board for Certified Counselors
3-D Terrace Way
Greensboro, NC 27403
(919) 547-0607

Source: American Counseling Association.

APPENDIX G

COMMONLY USED ABBREVIATIONS

@	about
A.A.	Alcoholics Anonymous
a.c.	before meals
A.C.S.W.	Academy of Certified Social Workers
ad. lib.	as much as desired
adm.	admitted
adol.	adolescent
AMA	against medical advice
amt.	amount
a.o.	anyone
appt.	appointment
as.	of each
ASA	against staff advice
asap	as soon as possible
ASR	at staff request
B/4	before
b/c	because
b.f.	boyfriend
b.i.d.	twice a day
b.i.n.	twice a night
bl.	black
B.M.	bowel movement
B.P. or B/P	blood pressure
bro.	brother
$\bar{c}$ or w/	with
c	canceled
Ca	calcium
CA	cancer
C.A.C.	Career Assessment Inventory
cap.	capsule
C.A.P.S.	Career Ability Placement Survey Test
C.B.C.	complete blood count
cc	cubic centimeter

cl or ct.	client
CNS	central nervous system
c/o	complains or complained of
C.P.T.	Common Procedural Terminology
C/R	cancel/reschedule
D or dos.	dose
D&A	drug and alcohol
D.A.T.	Differential Aptitude Test
d.c.	discontinue
Detox	detoxification
dis.	discontinued
D.N.A.	did not appear
dr. or z	dram
Dr.	doctor
D.S.M.	Diagnostic and Statistical Manual
D.T.'s	delirium tremens
Dx	diagnosis
E	evolved
ECT	electroconvulsive therapy
EEG	electroencephalogram
e.g.	for example
EKG or ECG	electrocardiogram
e.o.	everyone
ER, EW, or EMR	emergency room or emergency ward
exam.	examination
fa.	father
fl. or fld.	fluid
F.R.	fully resolved
ft.	feet
G.A.	Gamblers Anonymous
g.f.	girlfriend
G.I.	gastrointestinal
gm.	gram
gma.	grandmother
gmpa.	grandfather
gr.	grain
GRP	group
gtt.	drop
H & P	history and physical exam
h.s.	hour of sleep or bedtime
ht.	height
H2O	water
Hx	history
ICDM 10	International Classification of Diseases Manual (Vol. 10)
i.e.	that is
Iden.	identification
I.M.	intramuscular
ind.	individual

info.	information
I.S.B.	Incomplete Sentences Blank
I.V.	intravenous
J.C.A.H.O.	Joint Commission on Accreditation of Healthcare Organizations
K	potassium
L.	left
lge.	large
liq.	liquids
L.P.C.	Licensed Professional Counselor
L.P.C.C.	Licensed Professional Clinical Counselor
L.P.N.	Licensed Practical Nurse
mcg.	microgram
m Eq.	milliequivalent
Mg	magnesium
mg.	milligram
Mg SO4	magnesium sulfate
min.	minute
ml.	milliliter
mm.	millimeter
M.M.P.I.	Minnesota Multiphasic Personality Inventory
mo.	mother
mod.	moderate
mot.	motivation
MS	mental status
Mtg.	meeting
N.A.	Narcotics Anonymous
NA	not applicable
Na.	sodium
n/ach	not achieved
N.C.C.	National Certified Counselor
neg.	negative
n.o.	no one
no. or #	number
ns	no show
NPO	nothing by mouth
N & V	nausea and vomiting
obs.	observation
O.B.S.	organic brain syndrome
o.d.	everyday
O.D.	overdose
oint.	ointment
o.m.	every morning
OTO	one time only
oz. or z	ounce
p.	present
p.c.	after meals
pil.	pill

P.O.	probation officer
p.o.	by mouth
P.R.	partially resolved
prn	whenever necessary, as needed
pt.	patient
q.	every
q.d.	every day
q.h.	every hour
q.i.d.	four times a day
q.o.	every other
q.o.d.	every other day
q.s.	quantity sufficient
q.2h.	every two hours
q.4h.	every four hours
R. or R	right
re:	regarding
rec.	received
R.N.	Registered Nurse
Rx	prescription, medication
s̄	without
s.c.	subcutaneously
schiz.	schizophrenia
sib.	sibling
sig.	label it
sis.	sister
S.O.	significant other
s.o.	someone
sol.	solution
s.o.s.	if necessary
ss	one half
stat.	immediately
tab.	tablet
tbsp.	tablespoon
T/C	telephone call
temp.	temperature
t.i.d.	three times a day
t.i.n.	three times a night
T.O.	telephone order
T.P.R.	temperature, pulse, respiration
tr.	tincture
tsp.	teaspoon
Tx	treatment
U.	unit
U.R.	utilization review
UR	unresolved
V.O.	verbal order
w (or c̄)	with
w/i	within

wky.	weekly
w/o	without
wt.	weight
×	times
∴	therefore
↑	increase
↓	decrease
♂	male
♀	female

APPENDIX H

SAMPLE INTERN EVALUATION FORM

Intern's name _____

Date of this evaluation _____ Name of agency _____

Univ. supervisor _____ Agency supervisor _____

Title _____ Title _____

Professional degree _____ Professional degree _____

Licensed as _____ No. _____ Licensed as _____ No. _____

Certified as _____ No. _____ Certified as _____ No. _____

Internship dates __/__/__ to __/__/__ No. yr previous pd. m.h. work _____

Intern paid? Yes _____ No _____ Agency _____

Total no. placement hrs _____ Address _____

Suggested Competencies for Interns

4 = Outstanding   3 = Good   2 = Fair   1 = Poor   NA = Not applicable

I. Communication skills

   a. Verbal skills ____

   b. Writing skills ____

   c. Knowledge of nomenclature ____

Comments:

II. Interviewing

    a. Structure of interview ____

    b. Attending behaviors ____

    c. Active listening skills ____

    d. Professional attitude ____

    e. Interviewing techniques ____

    f. Mental status evaluation ____

    g. Psychosocial history ____

    h. Observation ____

    i. Use of questions ____

    j. Reflection ____

    k. Empathy ____

    l. Respect for differences ____

Comments:

III. Diagnosis

    a. Knowledge of assessment instruments ____

    b. Knowledge of current DSM ____

    c. Use of records ____

    d. Ability to formulate a preliminary diagnosis ____

Comments:

IV. Treatment

    a. Ability to draw up a treatment plan ____

    b. Ability to perform individual counseling ____

    c. Ability to perform marital counseling ____

    d. Ability to perform conjoint counseling ____

    e. Ability to perform family counseling ____

    f. Ability to perform group counseling ____

    g. Crisis intervention skills ____

h. Ability to deal with various populations _____

i. Ability to make progress notes _____

Comments:

V. Case management

    a. Knowledge of agency programs and professional staff roles _____

    b. Knowledge of community resources _____

    c. Discharge planning _____

    d. Follow-up _____

    e. Record keeping of client management _____

    Comments:

VI. Agency operations and administration

    a. Knowledge of agency mission and structure _____

    b. Awareness of roles of administrative staff _____

    c. Knowledge of agency goals _____

    d. Understanding of agency care standards _____

    Comments:

VII. Professional orientation

    a. Knowledge of counselor ethical codes _____

    b. Knowledge of agency professional policies _____

    c. Ability of intern to seek and accept supervision _____

    Comments:

Please write a brief summary statement of the intern as a future counselor.

| **Intern** | **Agency Supervisor** | **Univ. Supervisor** |

INDEX

TO THE OWNER OF THIS BOOK

We hope that you have found *The Counselor Intern's Handbook* useful. So that this book can be improved in a future edition, would you take the time to complete this sheet and return it? Thank you.

School and address: —————————————————————————

Department: —————————————————————————————

Instructor's name: ————————————————————————————

1. What I like most about this book is: ————————————————

——————————————————————————————————————

——————————————————————————————————————

2. What I like least about this book is: ————————————————

——————————————————————————————————————

——————————————————————————————————————

3. My general reaction to this book is: ————————————————

——————————————————————————————————————

4. The name of the course in which I used this book is: ————————

——————————————————————————————————————

5. Were all of the chapters of the book assigned for you to read? ————

   If not, which ones weren't? ————————————————————

6. In the space below, or on a separate sheet of paper, please write specific suggestions for improving this book and anything else you'd care to share about your experience in using the book.

——————————————————————————————————————

——————————————————————————————————————

——————————————————————————————————————

——————————————————————————————————————

——————————————————————————————————————

Optional

Your name: _____ Date: _____

May Brooks/Cole quote you, either in promotion for *The Counselor Intern's Handbook* or in future publishing ventures?

Yes: _____ No: _____

Sincerely,

Chris Faiver
Sheri Eisengart
Ronald Colonna

FOLD HERE

FOLD HERE

Brooks/Cole Publishing is dedicated to publishing quality books for the helping professions. If you would like to learn more about our publications, please use this mailer to request our catalogue.

Name: _____

Street Address: _____

City, State, and Zip: _____

Laurie Burt— lburthome@aol.com   617/965-4279
              lburt@fhe.com   617/832-1111